ENGINEERING STATISTICS
A CONCEPTUAL APPROACH
STA 381

William Rayens

VAN-GRINER
LEARNING

Engineering Statistics

STA 381
A Conceptual Approach

Copyright © by William Rayens

Photos and other illustrations are owned by Van-Griner or used under license.

About the Cover Design: The artwork on the cover was done by Jamie Van Landuyt, a studio artist working in Cincinnati. Just as her abstract art involves layering and obscured compositional elements, so do statistics as numbers belie the underlying conceptual complexity.

Printed in the United States of America
10 9 8 7 6 5 4 3 2
ISBN: 978-1-61740-755-0

Van-Griner Publishing
Cincinnati, Ohio
www.van-griner.com

President: Dreis Van Landuyt
Project Manager: Brenda Schwieterman
Customer Care Lead: Lauren Houseworth

Rayens 755-0 Su19
312621-316622
Copyright © 2020

Acknowledgments

The author would like to thank all the University of Kentucky teaching assistants and primary instructors who were brave enough to use this book from its very beginnings! Extra special thanks go out to William Griffith, Melissa Pittard, Pat Cain, Amanda Ellis, Nathan Koebcke, Dustin Lueker, Jeff Modenbach, and Gabby Ness, who recognized the wisdom in Dr. Rayens' methods and worked tirelessly to create an environment for shared learning. Finally, the author would like to thank Peggy Saunier and Aaron Profitt for reviewing this workbook in previous editions and offering invaluable suggestions to help it evolve.

Table of Contents

MODULE 1

Human Inference 1

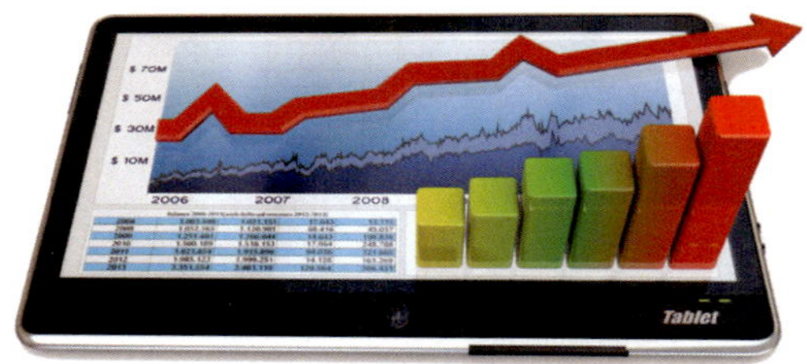

MODULE 2
Confidence Intervals 35

MODULE 3
Formal Inference 61

Introduction

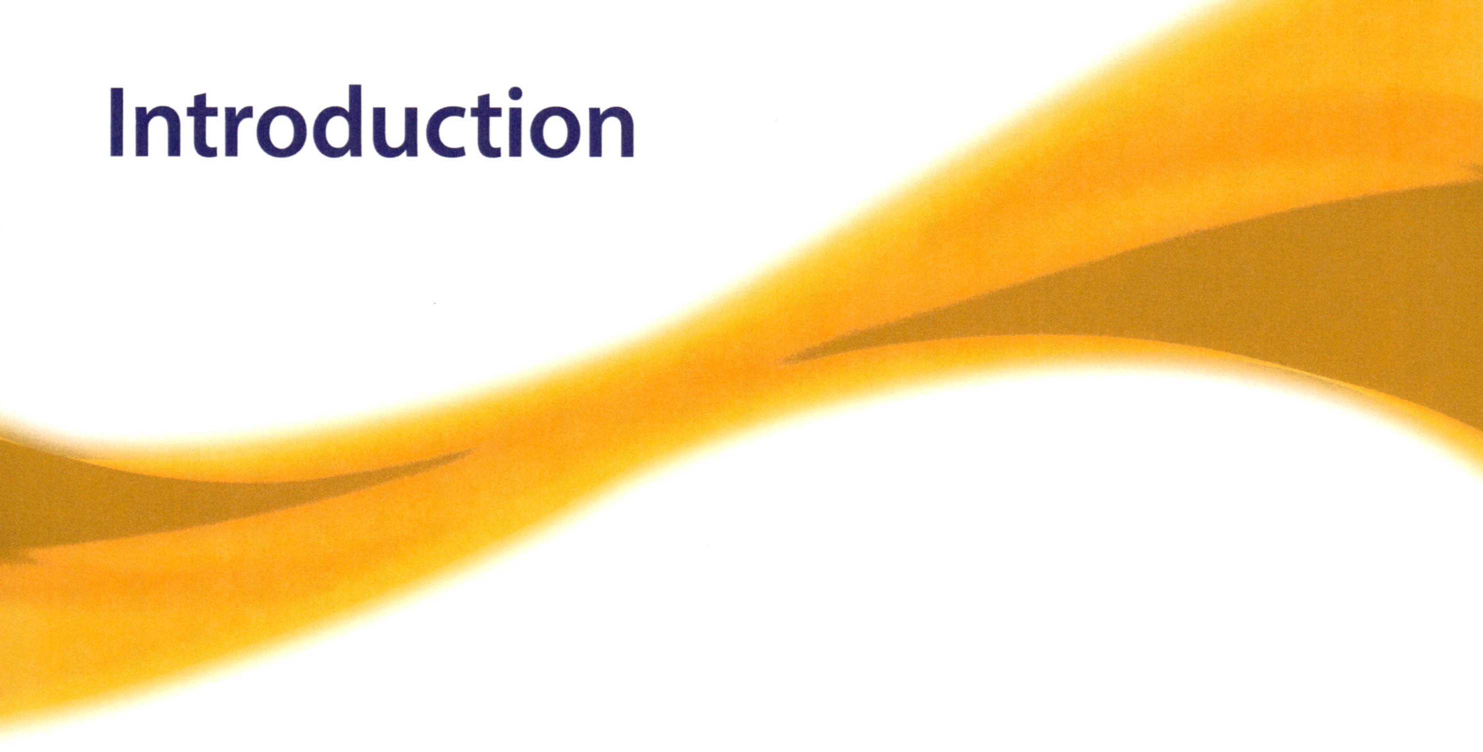

Beyond the Numbers
Student-Centered Activities for Learning Statistical Reasoning

Overview of the Workbook

Dr. William Rayens has taught statistical reasoning at the University of Kentucky (UK) for more than thirty years. This workbook addresses the needs of the statistical reasoning community, as well as the ever-growing universal need for statistically-literate citizens. There are several conceptually-focused statistical reasoning books on the market, but *Beyond the Numbers: Student-Centered Activities for Learning Statistical Reasoning* is distinct in three noteworthy ways:

1. **This text is unambiguously focused on statistical reasoning.** Our goal is to help students competently consume statistical ideas that meet them where they live—both academically and personally. If you measure worth by the number of distinct methods covered, then this workbook might seem light. If you measure worth based on the weight of the statistical reasoning that the carefully chosen methods surface, then the workbook is heavy, indeed. Computations and exposure to different methods are still valued, since they surely increase students' life skills. But we have been careful not to allow students to be distracted by computations, since it has been our experience that they often find those easier to grasp than the more important understanding of what is gained from such computations.

2. **This text is designed to shift the primary responsibility of learning to the student.** Borrowing the words of STEM (science, technology, engineering, and mathematics) educator Robert Talbert, the goal of this shift is to change the mindset of the student from a "renter" to an "owner." In a classroom of renters, fees are paid and the management is expected to deliver. In a classroom of owners, however, students realize that it is their responsibility to engage, absorb, and retain. The instructor's very difficult (and very important) job, then, is to prepare an environment in which that can happen. This workbook is designed specifically to facilitate students' transition from renters to owners. How so? It does not contain much stand-alone content, but instead guides students through carefully curated activities, most of which require independent reading, web-based research, and/or collaboration. Methodological and conceptual content is presented at appropriate places in Read All About It sections that are seamlessly interspersed with the other activities. The same content is also available on student-friendly, ADA-compliant videos, which are freely available on YouTube. In a self-driven environment carefully crafted by the instructor, the organization and scope of the workbook provide the coherence and direction needed to do the work and learn the material. And make no mistake—it is indeed the student doing the work and learning the material.

3. **Finally, this text's pedagogy is not mathematically driven.** This was perhaps the hardest safety net for Dr. Rayens to let go of. However, having taught literally thousands of students over the years, Dr. Rayens was confident that students taking an entry-level course—especially a concepts-focused course—could not fully appreciate or benefit from the way that most statistics books are organized. Though some attempt to hide it, statistics texts unfold along a path that is defined by the mathematical development of the material. It is not enough simply to hide the mathematics—the pedagogy should follow the path that is most conducive to student learning, not the one dictated by mathematical development. This is the same pedagogical approach taken by Professor Xiao-Li Meng when he created the innovative course "Real-Life Statistics: Your Chance for Happiness (or Misery)" at Harvard.

One could say that this workbook is the product of a serious academic midlife crisis shared among Dr. Rayens and several of his colleagues! It is perhaps more correct to say, however, that this text is the product of clarity of vision that comes only after spending decades in the classroom. **The guiding directive is clear: Students have to become more involved in their learning, and professors have to be willing to reassess their roles in order to facilitate that shift. The need is there—this workbook was created simply as a response to the need.**

How This Workbook Is Laid Out

Beyond the Numbers: Student-Centered Activities for Learning Statistical Reasoning has undergone several important revisions over the years. Dr. Rayens has been careful not to change too much (or too drastically), but at the same time, to incorporate user feedback and optimize the existing text. Broadly speaking, the goal of this workbook is to help students understand the inferential arguments they encounter as part of their daily lives. Most of these daily encounters occur in the forms of polls and surveys; social and medical experiments; and the human inference from simple statistical constructs like tables and graphs. Like previous editions, this one is divided into three major modules to reflect these sources:

1. Human Inference

2. Confidence Intervals

3. Hypothesis Testing

For ease of use, we open every module with learning outcomes. These outcomes are closely tied to the content in the module. Within each of the modules the content is organized in three ways:

CONTENT

Content to support the major concepts in the workbook are made available two ways: directly in the workbook in compact Read All About It activities, and on corresponding videos. You are more than welcome to show the videos; they can be found at **www.statconcepts.com/videos.** You may opt to deliver this material differently; however, to make the workbook work most easily we recommend the videos. The content material is summarized on page XIV.

BEYOND THE NUMBERS (BN)

This broad category contains a variety of activities. Beyond the Numbers activities are designed more as content packets than as traditional exercises. While they do provide needed practice and reflect on activities previously completed, more often than not they surface new concepts related to a particular topic addressed in the module. They can be assigned in class or as homework, and you have the option to select the ones that best fit your course. Each BN covers specific learning outcomes.

BEYOND THE CLASS (BC)

At the end of each module you will find the Beyond the Class activities. They are designed to both keep the student focused on the overarching goals of the course and keep them connected with the world around them. For example, a student may be asked to comment on a media article that you assign; you may ask them to provide their own article and to write about it; or to complete a conceptually-focused exercise using real data. It is imperative to connect what they do *inside* the classroom with what goes on *outside* the classroom. Again, we have designed these with flexibility in mind, and you may assign them all, a few or none.

CAPSTONES

Two larger-scale projects are included at the end of Module 3. Both are designed to capture as much of what students have learned from all three modules as possible. But beyond that, the capstones are designed to give students an opportunity to research a topic of their choice and create their own hypothesis. In the first capstone students will collect their own data by way of a survey, and in the second students are directed to design and implement an experiment. The purpose of both assignments is to apply what the student has learned in the course, then present and interpret the results in a formal paper. Both Capstones have been designed for either individual or group assignments.

Pedagogy

Activities are grouped together in a natural order, providing an experience that is gently structured, yet flexible. This three-stage organization both maintains a high intellectual bar for students and makes teaching from the workbook significantly ($p < 0.05$) easier than it might otherwise be.

The sequencing works like this. Our first goal is to **Engage** the student in the key point to be made. Let's take a part of Module 3 devoted to understanding hypothesis testing in the media as an example. As we morph into this part of Module 3, we are leaving a substantive treatment of sensitivity and specificity. The key deliverable is for the student to be able to analyze the ubiquitous phrase "statistically significant" in a real context, and say clearly what that means. They should be able to identify what comparisons were being made, how they turned out, and what risk was involved in the decision that was made.

Once this primary deliverables has been engaged and practiced, we then **Reflect** on what we have done, and perhaps not done. In the example being discussed, it is natural to then turn to a discussion of practical significance. We then revisit some cases where statistical significance was clear, but why anyone would care is not. And we may choose to use software to look at the role of sample size in a distortion of statistical importance.

Following the reflection we have the opportunity to **Extend** what we have just learned, if the instructor chooses. For instance, in the example being discussed, we then have the option of extending our knowledge to the real difference between failing to reject H_0, and accepting it, what power is and how sample size affects it, and even to the philosophically difficult topic of error rates versus p-values, which is probably the most tragically misunderstood topic in statistical science. These reflections and extensions are all brief and to the point. They are designed to make the student statistically more aware, but not designed to turn her into a full professor of statistics. Some of the extensions throughout the workbook are intellectually quite challenging.

Relevancy

If you are going to ask students to be actively involved in their own learning, it is important you *engage* them. With examples that are relevant to their day-to-day encounters students are more likely to relate to the material. Once the connection is made students tend to be drawn into the content and consequently become more involved in their own learning. To help make these critical connections, the author has been careful in his choice of activities. Throughout the workbook he addresses current topics that are not only illustrative in content, but are also relatable to the student. Some examples include:

- A February 2014 study on gay marriage to facilitate the construction and interpretation of a confidence interval, when the margin of error is given.

- A 2014 Gallup Organization–Purdue University report on does it matter where you go to college to introduce the idea of using confidence intervals to make hypothesis-testing like decisions.

- A study on the efficacy of social networking systems as instructional tools to demonstrate the empirical rule.

Another key to engagement is the presentation of the material. We have deliberately chosen a design that is more open, friendly and less formal than other textbooks that cover similar material. Don't be fooled, the material is there, we have opted for a format that is more student-appealing.

Flexibility and the Flipped Classroom

In a student-centered environment, instructor flexibility is not a problem. If anything, Dr. Rayens provided *too much* flexibility during the first couple of years developing this workbook. As mentioned previously, the instructor has the pleasure (and the burden) of deciding how to organize the classroom environment around the workbook activities. This is possible because the workbook was designed with flexibility in mind. Instructors are able to select a path through the workbook, and then move through the activities in a way that fits their unique teaching styles and course structures. For example, you only need one simple hypothesis about a proportion to teach the underlying logic of hypothesis testing. However, if time permits and the students' backgrounds allow, it would also be beneficial to introduce tests involving both one and two means. The deliverable is still conceptual, but there are more chances to increase students' life skills along the way. Either route is possible with this workbook. This may seem like a challenge, but it is perhaps the truest teaching you will ever engage in.

At the University of Kentucky, the lure of the flipped classroom was one of the most influential drivers of the flexibility embedded into this workbook. Flexibility allowed Dr. Rayens to move most of his scaled-down lecture content out of the classroom, transforming his classroom into more of a laboratory than a lecture hall. Inversion is not required to use this workbook, however. It is ideal for any type of classroom. In fact, most of the instructors who use it do not teach in a fully-flipped classroom. No matter the classroom organization, it is the student's participation in his own learning that is held as an unshakeable value among workbook users. *That* is the primary constant facilitated by this workbook.

How to Migrate to a Workbook

The two biggest challenges in moving to a workbook-centered classroom are, first, becoming comfortable with the reduced content, and second, adjusting to the new role you play in the course pedagogy. There is genuine comfort in assigning a reading out of a traditional textbook, repeating that content in a classroom presentation, and then assigning problems for students to complete on their own. In a workbook-centered environment, content is still important—it is simply delivered differently. The emphasis each day is not on a lecture, but on working through real-world activities or responding to media excerpts (a very humble version of what Professor Meng sought to accomplish at Harvard).

The language and tools needed to do this are provided in two ways. Most of the content is confined to the focused Read All About It sections of the workbook (alternatively, also on student-friendly, ADA-compliant videos that are available to anyone who needs them). Other content surfaces in the workbook as needed, often in the form of a brief preface to the activity that utilizes it.

Dr. Rayens and his colleagues all facilitate their workbook environments differently. If an instructor has 75 students, she may assign the content in advance, test the students on that content when they enter class, and then spend the rest of the class time working through exercises from the workbook. She may also take time at the end of class to have students report their findings. Dr. Rayens utilizes a similar dynamic in his large classes, though the implementation is more challenging because an increased class size might mean being located in an old-fashioned lecture hall with fixed theater-style seating. To accommodate these less-than-ideal conditions, Dr. Rayens assigns both the content and the workbook exercises in advance of class. When students arrive to class, Dr. Rayens typically has a large-class dynamic prepared to allow for a deeper look at the exercises due that day. Dynamics may involve clickers, discussions arising from short skits performed by the instructor and his teaching assistants, students presenting their findings for extra credit, or yes, even a five-minute PowerPoint presentation! As a workbook adopter, you will need to decide what format works best for you and your students. No matter what format you follow, the goal is the same—to encourage students to engage with the activities.

Incorporating Math into Your Course

 Touched on earlier in the preface (Overview of the Workbook), we think it is important to revisit the topic of *how* we incorporate math into the workbook. You will see throughout the workbook that we don't avoid the idea of having students work through formulas and apply their math skills; we just don't use math as the workbook's compass. Nor do we allow students to be tricked into thinking that math is what makes an elementary statistics course challenging. We don't believe that. The concepts are typically much more challenging than the math at this level. If you are specifically looking to assign material from the workbook that requires students to apply their mathematical skills, we have an answer for you. In this edition, when you see the icon above in the Beyond the Numbers header you will know that students, to solve some of the problems in the Beyond the Numbers, will need to work through some computations.

Software Options

You asked, and we answered! *Beyond the Numbers: Student-Centered Activities for Learning Statistical Reasoning* not only introduces datasets that instructors can use with software packages of their choosing—it also includes brand new activities designed specifically around those data sets. All students, from liberal arts majors to engineers, need to know how to use basic software suites to manipulate numbers, perform calculations, and create graphs. Sobered by how few UK students learned these basic skills as they progressed through college, Dr. Rayens endeavored to integrate them into this workbook. Dr. Rayens' concepts-based statistics course, it turned out, proved the perfect environment to help students develop these life skills. The datasets are not proprietary and can be used with any of the major software pages, including Microsoft Excel and Apple Numbers.

Large and Small Classroom Sizes—They Both Work

At the author's school, this course is taught in rooms with more than 135 students, rooms with approximately 75 students, and even an honors section with approximately 30 students. While group work and lively student interaction decrease as the class size increases, the course can be tailored to reach any class size effectively. The goal of the workbook is to increase a student's participation in his own learning. This does not necessarily mean that students need to huddle up or take turns speaking in front of the class. Rather, it means that they should be assigned exercises that require them to engage, reflect, and if possible, extend their understanding beyond the original exercise. Dr. Rayens and his colleagues have experimented with many different ways to effectively shrink a large classroom. A section of the workbook website is set aside for sharing their experiences.

Customize to Fit Your Course

Because each Beyond the Numbers and Beyond the Class is a standalone activity, you can easily select the ones that meet the needs of your course. You won't have to worry about missing content or creating a book that is "choppy" that your students find difficult to follow. Simplifying your selections, we have also categorized Beyond the Numbers into three groups: Engage, Reflect, and Extend. With several options in each group you are sure to find the one that best matches your teaching needs. If you are looking to add your own content, we can do that for you as well. We will seamlessly integrate your materials into the book to give them the same look and feel as the original text. Finally, if you are looking for a lower price, customization could be an excellent option. Already a low price we can make it lower by reducing the page count. If you are interested in any of our custom options contact help@van-griner.com.

Instructor Response to This Workbook

Dr. Rayens produced course videos, developed classroom activities, created student-centered dynamics, elevated the presence of technology in the classroom, redesigned recitations to fit the skills of teaching assistants, created training sessions for instructors, and wrote the first edition of this workbook—all in same compressed time period. Even though the first several editions of this text reflected the dizzying conditions under which it was written, Dr. Rayens' adopters were generally very pleased with the workbook and the course for which it was created. Some even went so far as to suggest that this may have been the most honest *teaching* they had done in years—maybe ever. It quickly became apparent that a class built around workbook activities forced students first to attend class, and second to take a more active role in their learning. It also became apparent, however, that students do not always embrace new opportunities for independence. In fact, independence is a foreign and sometimes uncomfortable concept for many students.

With the help of his instructor colleagues and student feedback, Dr. Rayens began adjusting the workbook to better fit students' needs. The first iteration had too little structure and too few activities, but each new revision improved upon the last. Often feeling like he was creating a whole new course for each semester he taught, Dr. Rayens tested assembling workbook activities into a syllabus in countless different ways. The freedom was both great and, at times, daunting. Throughout this process, Dr. Rayens dedicated himself to listening to the experiences of his adopters. These shared experiences, collected from teaching assistants to tenured faculty, dramatically affected the workbook and course as they exist today. As it has matured, expanded, and been reorganized, this workbook has come to allow an instructor to shift easily from a sage-on-the-stage dynamic to one that promotes student involvement. And it seems to matter very little how inverted the classroom is. What matters is developing the correct environment for learning to take place and for students to shoulder their share of that effort. This workbook can help immensely in achieving that goal.

Teaching Materials

The development of this workbook has proven an evolutionary experience, as each new edition has improved upon the last. Dr. Rayens has listened and learned, and with this edition, he is excited to introduce a brand new companion website (**www.statconcepts.com**) where you will find teaching ancillaries and other materials to help you transform your classroom into a student-driven learning environment. On this website, you will have access to:

- Data sets referenced in the workbook

- Videos that provide a verbal presentation of the Read All About It content

- Teaching ideas

    a. Exam questions

    b. Sample syllabi

    c. Sample daily schedules

Content Summaries

Statistics educators have long known that ordinary statistical concepts can be extraordinarily hard for students to grasp. The underlying explanations are often abstract, and at their core, mathematically complex. Unfortunately, it can be difficult to understand even common concepts like margin of error and statistical significance without some successful encounters with the role played by the underlying mathematics. This workbook is intended to facilitate these kinds of encounters, but supporting content has to come from somewhere. For this workbook the content is confined to the carefully focused Read All About It sections, and alternatively placed on student-friendly, ADA-compliant videos that are freely available at **www.statconcepts.com.** This content is summarized below.

Module 1—Human Inference

NUMBER SENSE—BASIC NUMERACY

Basic numeracy is a critical step toward the goal of correctly forming human inferences from statistical constructs. In this content, we take a look at decimal point errors and deceptive graphs, among other challenges to basic numeracy.

NUMBER SENSE—COMPUTATIONS AND BENCHMARKS

Competence with fractions, percentages, and common benchmarks goes a long way toward the goal of correctly forming human inferences from statistical constructs. In this content, we look at examples of all of these concepts.

CONFOUNDING AND THE LANGUAGE OF EXPERIMENTS—INTRODUCTION

Credible inferences from experimental data have to be free from confounding. In this content, we introduce two primary sources of confounding: lack of proper comparison and improper randomization of subjects to treatments.

CONFOUNDING AND THE LANGUAGE OF EXPERIMENTS—PROPER COMPARISONS AND RANDOMIZATION

Both the placebo effect and lack of randomization can create obstacles to making credible inferences from experimental data. In this content, we look at examples of both.

CONFOUNDING AND THE LANGUAGE OF EXPERIMENTS—STATISTICAL SIGNIFICANCE

Credible inferences from experimental data have to be held to a formal standard of statistical significance. In this content, we introduce this concept in a non-mathematical context.

CORRELATION AND CAUSATION—SCATTERPLOTS

A simple scatterplot is an informal, yet useful visual means of addressing both the direction and strength of a relationship between two variables. In this content, we define the relevant terms and look at examples.

CORRELATION AND CAUSATION—CORRELATION COEFFICIENT

The correlation coefficient is the most common numerical measure of the strength of a relationship between two variables that can represented by a scatterplot. In this content, we learn how to compute and interpret the correlation coefficient, and we look at a variety of examples.

CORRELATION AND CAUSATION—CAUSATION

Association doesn't imply causation because other variables may be influencing the relationship. However, correlation could indeed be evidence of causation. In this content, we revisit this persistent misunderstanding by way of explanations and examples.

Module 2—Confidence Intervals

SAMPLING—INTRODUCTION

Avoiding biased samples is good common sense, whereas reaching meaningful conclusions from proper samples is the business of statistical science. In this content, we look at some common problems with biased samples as we distinguish common sense from mathematical science.

SAMPLING—LANGUAGE AND TECHNIQUE

Collecting a probabilistic sample (like a simple random sample) is a critical step in formally estimating a population parameter with a sample statistic. In this content, we introduce the language needed to understand statistical sampling.

SAMPLING—CONFIDENCE INTERVALS

Simple formulas are available for calculating the margin of error and associated confidence intervals provided the data were collected in a simple random sample (or an equally statistically-correct fashion). In this content, we discuss these formulas, along with a very precise interpretation of confidence intervals.

SAMPLING—WHEN MOE DOESN'T APPLY

Non-sampling errors can substantially detract from the accuracy of survey data and are not addressed by the margin of error. In this content, we discuss the worrisome prevalence of non-sampling errors, how one major polling organization has reacted, and what you can do when faced with such errors.

Module 3—Formal Inference

SENSITIVITY AND SPECIFICITY—INTRODUCTION

False positive rates and false negative rates are commonly used to assess the integrity of a screening test. In this content, we introduce both, along with their simple connections to specificity and sensitivity.

SENSITIVITY AND SPECIFICITY—COMPUTATIONS AND EXAMPLES

Simple fractions are used to compute sensitivity and specificity in situations whereby test results and actual data are arrayed in 2 × 2 tables. In this content, we look at several examples and practice these computations.

HYPOTHESIS TESTING—AS A DIAGNOSTIC TOOL

It can be useful to view statistical hypothesis testing as a kind of screening test. In this content, we develop the vocabulary to make explicit that correspondence and look at several examples.

HYPOTHESIS TESTING—APPLYING THE CONCEPTS

To understand how the term "statistical significance" is being used in the media, you must ask what is being compared, what the null and alternatives are, how the comparison turned out, and what risks were involved in the decision that was made. In this content, we carefully address the interpretation of this phrase using real examples.

HYPOTHESIS TESTING—PRACTICAL SIGNIFICANCE

Statistical significance is a mathematical tool used to assess whether treatment differences are likely to have occurred by chance. Practical significance addresses whether the observed difference is big enough to care about. In this content, we highlight these differences by way of discussion and examples.

HYPOTHESIS TESTING—COMPUTATIONS

Testing a simple hypothesis is a process involving the computation of a standard score. This score is then taken to a table to identify the p-value needed to make the final decision about whether the null will be rejected. In this content, we discuss and illustrate these steps. We are careful to distinguish Type I error rates from the well-known p-value.

Grading Options—Using Peer Review

One emphasis of this workbook is student response. In the Beyond the Class and Beyond the Numbers exercises, the student is as often asked to explain an answer as well as perform a computation. While these short answer responses are immensely beneficial to student learning, they also tend to create additional grading burdens for instructors. To meet this challenge, and to increase the time students spend with the exercises, Dr. Rayens, his colleagues, and faculty across the country often use peer review tools. Dr. Rayens has used and can recommend two options: Canvas Peer Review and a software tool out of UCLA called Calibrated Peer Review (CPR).

CALIBRATED PEER REVIEW

Calibrated Peer Review, often just referred to as CPR, is a peer review utility created by professors at the University of California, Los Angeles. This tool effectively leaves the grading of assignments in the hands of other students, but does so in a way that is arguably even more useful to students than the instructor marking the assignments herself. For each assignment, students are electronically calibrated, assigned peer documents to grade, and then required to grade their own assignments as well. Each step of this process is scored automatically as a part of the overall assignment grade. All calibration materials and rubrics are built into the assignments and are made available on the sponsor site. All the Beyond the Numbers and Beyond the Class activities in this workbook are fully developed on the central CPR site.

There is a learning curve for instructors and students to familiarize themselves with CPR, and some students struggle initially with the complexity, but the payback on that time investment can be excellent. For more information about CPR, please visit **http://cpr.molsci.ucla.edu/Home.aspx** or contact Dr. Rayens directly to learn how it is being used with this workbook in particular.

Canvas is a well-known learning management system (LMS) by Instructure. It has recently become the official LMS for the University of Kentucky and Dr. Rayens has experimented in depth with using the peer review tool that is built into that system. While this tool has far fewer bells and whistles than CPR, it is much easier for students and instructors to use. With the Canvas tool an instructor can designate a Beyond the Numbers assignment as "peer graded" and the system can be prompted to randomly assign one or more submissions to each student to be peer reviewed, or these assignments can be made manually. Students do the peer grading from within Canvas, using rubrics that accompany each assignment. Canvas rubrics have been designed for each assignment in the workbook and can be made available from Dr. Rayens upon request.

There is far less of a learning curve for instructors and students with the Canvas tool. But you can't do nearly as much as within CPR. As a way to force an encounter between the student and a rubric, to effectively communicate what was valued about an assignment, it works well. For more information about the Canvas Peer Review Tool, please visit **https://guides.instructure.com/m/4152/l/54366-what-is-a-peer-review-assignment** or contact Dr. Rayens directly to learn how it is being used with this workbook in particular.

Is This Workbook a Good Fit for You?

If you have read this far, you are probably considering this book—you might already think that it's a good fit for you. To be honest, you probably already know if you are ready to make the change. If you feel that it's time to implement a concepts-driven course structure that engages students and imparts lasting statistical reasoning skills, then this workbook is an excellent way to do so. You may be asking yourself, "If students are enough engaged in their own learning, what do they need me for?" That's a challenging question, but one that deserves to be asked (though maybe in private, among other instructors)! According to Dr. Rayens, students need instructors to prepare an environment that is right for learning; to provide essential, low-profile navigation through that environment; and to gently force a shared responsibility for the learning taking place. When it works, students tend to conclude "I did this all by myself." You know better, of course, but you can nevertheless celebrate the intellectual growth that has just taken place. Candidates who have had similar reflections will have the best success with this workbook.

If you work at a large-enrollment school, you may use teaching assistants to lead classroom discussions (as Dr. Rayens does at UK). If so, you may be wondering if teaching assistants can be relied upon to create a learning environment rather than simply lecture on the material. In fact, we have found that teaching assistants are some of the best candidates to adopt this workbook. This is true partly because they are unencumbered with legacy teaching methods, and partly because they are typically younger and more open to trying new things. At UK, all first-year teaching assistants are required to apprentice with an experienced instructor for a full academic year before taking on this (or any) course alone. So while it may not be advisable for a first-year student to teach this workbook by themselves, teaching assistants have excelled in doing so once the apprentice year is complete. In fact, new revisions have made it even easier for seasoned teaching assistants to teach this workbook: a revised organizational structure frames how the semester pedagogy naturally flows, and the two-fold increase in activities reduces (or eliminates) the need for teaching assistants to provide their own supplemental materials.

About the Author

Dr. William Rayens is Professor of Statistics, Director of Educational Initiatives, and Associate Chairperson in the Department of Statistics at the University of Kentucky. Rayens has an extensive research record focused primarily on the development of multivariate and multi-way statistical methodologies mostly related to problems in chemistry and the neurosciences. He has mentored several Ph.D. students and has been honored at both the College and the University level as an outstanding teacher. Rayens also served as Assistant Provost for General Education during which time he was tasked with implementing new general education reforms at the University of Kentucky, the first changes to that program in almost 30 years. Rayens created STA 210: Introduction to Statistical Reasoning in 2010, and designed the one-of-a-kind Technologically Enhanced Active Learning rooms in the Jacobs Science Building where he and his instructors are privileged to teach.

Contact the Author

Thank you for reviewing this text! Countless hours have been devoted to this workbook—both by Dr. Rayens and by the colleagues whose insights and experiences inspired him. *Beyond the Numbers: Student-Centered Activities for Learning Statistical Reasoning* would not be what it is today without the considerable input of instructors at UK and across the country. As such, Dr. Rayens welcomes you to submit questions and feedback about the workbook, its accompanying videos, or website to rayens@uky.edu. As committed as he is to improving the workbook, Dr. Rayens is equally committed to helping you implement it in your own classroom. If you have any questions about Dr. Rayens' process or the course he created, feel free to send them along as well!

Human Inference

Overarching Goal

The primary intent of this module is to develop the skills needed to absorb common statistical information and to correctly form the associated human inferences.

Learning Outcomes

You'll know you have successfully completed this module when you are able to:

1. **Identify** categorically good or bad statistical summaries, charts, and graphs, and **explain** the reasons they are so categorized.

2. **Identify** categorically good or bad statistical arguments based on statistical summaries, charts, and graphs, and **explain** the reasons they are so categorized.

3. **Compute** basic statistical summaries and **create** simple graphs.

4. **Define** and **apply** basic experimental design vocabulary.

5. **Identify** confounding variables and **evaluate** their effects on experimental results.

6. **Explain** the role of randomization in simple experimental design.

7. **Explain** in non-mathematical terms the concept of statistical significance.

8. **Identify** and **assess** associations seen in scatterplots and two-way tables.

9. **Distinguish** the concepts of *association* and *causation,* and **explain** how they offer different types of evidence.

10. **Compute, apply,** and **interpret** the correlation coefficient.

Perceptions, Pictures, and Percents 

Name: ____   Section Number: ___________

To be graded, all assignments must be completed and submitted on the original book page.

Carefully read and think about each of the exhibits below. Then, give detailed answers to the questions that accompany each exhibit.

EXHIBIT 1

Dilbert

The following dialogue is from a Dilbert cartoon published April 17, 1996.

Secretary:	Oh my! This is shocking!
Boss:	What?
Secretary:	*40% of all sick days taken by our staff are Fridays and Mondays!*
Boss:	What kind of idiot do they think I am?
Secretary:	Not an idiot savant, they can do math.

Questions

1. What might you infer from the secretary's second statement (the one in italic)?

2. Explain what the secretary is trying to say with her last statement.

EXHIBIT 2

CD Sales

Percentage market share for three online sellers of compact discs.

Source: David S. Moore and William I. Notz, *Statistics: Concepts and Controversies,* 6th ed. (New York: W. H. Freeman and Company, 2006), 195.

Questions

1. Ignoring the numbers on the graph, what might you infer about the market share for CDNow compared to CUC?

2. How does your visual impression compare to what the numbers say?

EXHIBIT 3

Oral Contraception

In October 1995 the U.K. Committee on Safety of Medicines warned that third-generation oral contraceptive pills increased the likelihood of potentially life-threatening blood clots in the legs or lungs twofold—that is, by 100%. This information was passed on in "Dear Doctor" letters to 190,000 general practitioners, pharmacists, and directors of public health and in an emergency announcement to the media.

Questions

1. What might we infer about the safety of third-generation oral contraceptive pills from reading about this 100% increase?

2. Read the ABC News report from 2008 at http://abcnews.go.com/Technology/ Story?id=6034371&page=3. Summarize the discussion of relative risk versus absolute risk.

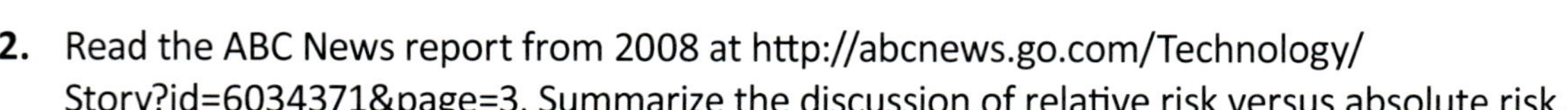

Computation and Common Sense

Name: ____   Section Number: __________

To be graded, all assignments must be completed and submitted on the original book page.

Carefully read and think about each of the exhibits below. Then, give detailed answers to the questions that accompany each exhibit.

EXHIBIT 1

Book Sales

Title: *Harry Potter and the Deathly Hallows* Sells 8.3 Million in First 24 Hours

Source: The Associated Press, July 22, 2007, 6:28 p.m.

It is the richest going-away party in history.

"Harry Potter and the Deathly Hallows," the seventh and final volume of J.K. Rowling's all-conquering fantasy series, sold a mountainous 8.3 million copies in its first 24 hours on sale in the United States, according to Scholastic Inc.

No other book, not even any of the six previous Potters, has been so desired, so quickly. "Deathly Hallows" averaged more than 300,000 copies in sales per hour—more than 50,000 a minute. At a list price of $34.99, it generated more than $250 million of revenue, more than triple the opening weekend take for the latest Potter movie, "Harry Potter and the Order of the Phoenix," which came out July 10.

"The excitement, anticipation, and just plain hysteria that came over the entire country this weekend was a bit like the Beatles' first visit to the U.S.," Scholastic president Lisa Holton said in a statement Sunday.

Question

1. We breathe about 15 times per minute, and a hummingbird flaps its wings about 3,000 times per minute. So a rate of 50,000 copies per minute would truly take our breath away and be faster than we could discern with our eyes. Is the 50,000 figure right? Explain.

Spousal Abuse

A letter to the editor of the *New York Times* complained about a *Times* editorial that said, "An American woman is beaten by her husband or boyfriend every 15 seconds." The writer of the letter complained: "at that rate, 21 million women would be beaten by their husbands or boyfriends every year. That is simply not the case." He then went on to cite the National Crime Victimization Survey (NCVS), which estimates about 254,000 cases of violence against women by husbands or boyfriends each year. The NCVS reports cases, not incidents. So a case for a year could refer to several incidents. There are lots of statistics in this problem that need sorting out.

Question

1. Is the letter writer correct to claim that the *Times* overstated the number of cases of domestic violence against women? Be sure to elaborate and reason very carefully using all the information you have in the problem.

Really Random Reasoning

Name: ____   Section Number: __________

To be graded, all assignments must be completed and submitted on the original book page.

EXHIBIT 1

Was Kobe Streaking? _______________________

Kobe Bryant's field goal shots for the first game of the 2001 Western Conference finals between the Los Angeles Lakers and the San Antonio Spurs are as follows:

O O O X X X X X X X O O X O O O (halftime) X X O X O O X X O X X O O O X O X O X X O

(O Missed shot, X = Hit shot)

The announcer's first-half comment included sentiments such as "What an awful start" after Bryant missed three in a row, "The basket is looking mighty big right now" after he hit five in a row, and "This is an unstoppable run" after he hit the seventh in a row (and coincidentally, right before the run did in fact stop).

Questions

1. Looking at Kobe's first game field goals, do you think one can argue he got a "hot hand" starting on his fourth field goal attempt and first basket? Why or why not?

2. For our purposes, define a "run" as two or more shots that turned out the same way. That is, two or more hits in a row, or two or more misses in a row, can be considered a run. How many runs were there in Kobe's first game, per the results above? Statisticians often use the number of runs (defined a little differently) to formally assess randomness.

You Are so Random. Or Maybe Not. 

Instructions: Get into groups as directed by your instructor. Every person in a group will have the same answers to this assignment, but for record keeping, you may be instructed to have every person fill out her own sheet with those common answers.

Your Instructor will facilitate the following sequence of events, but must be out of the room while they are taking place.

1. Every group of students will be assigned at random to one of two tasks.

2. One set of groups (Task 1) will each be given a die and asked to roll it 20 times. If your group is assigned Task 1, then you record the sequence of odds and evens in Table 1.3 below.

3. The other set of groups (Task 2) will each be asked to make up what they think would be a convincing random sequence of odd and even outcomes. If your group is assigned Task 2, then you record those answers in Table 1.3 below.

4. Do not indicate anywhere on this sheet what task your group was assigned.

5. Your Instructor will return to the room and proceed to guess with amazing accuracy which task was assigned to which group. Ahh, the power of the statistical mind. Or maybe not!

TABLE 1.3 Roll Number ...

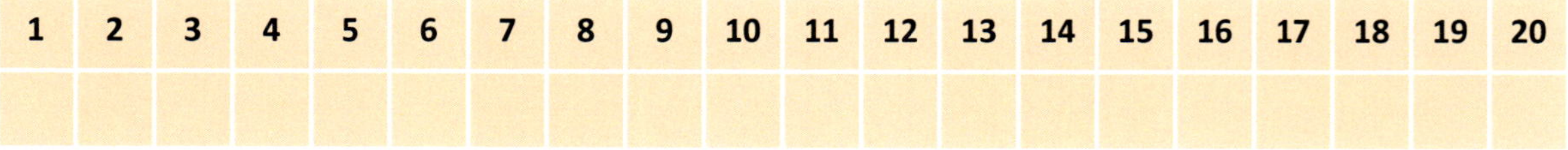

1	2	3	4	5	6	7	8	9	10	11	12	13	14	15	16	17	18	19	20

(Each of these 20 squares should have either an "O" or an "E" entered.)

Question

1. After experiencing this demonstration, what have you learned about the concept of randomness?

Hardwired to Slippery Thinking 

Name: ____   Section Number: __________

To be graded, all assignments must be completed and submitted on the original book page.

EXHIBIT 1

Cutting the Corpus Callosum

Answer the following questions while watching the "Split Brain" video at http://www.youtube.com/watch?v=ZMLzP1VCANo.

1. When Joe was shown the pan in his left visual field, what did he say he saw?

2. When Joe was shown the saw in his left visual field and a hammer in his right visual field, what did he say he saw?

3. When Joe was shown the saw in his left visual field and the hammer in his right, what was he able to draw with his left hand?

Have I Got a Story for You! ___________

Title: The Split Brain Revisited

Author: Michael S. Gazzaniga

Source: *Scientific American* 279, no. 1 (1998): 50–55

About the Study The author, psychology professor Dr. Michael Gazzaniga, describes part of the split brain study he conducted. On page 53 he writes:

> My colleagues and I studied this phenomenon by testing the narrative ability of the left hemisphere. Each hemisphere was shown four small pictures, one of which related to a larger picture also presented to that hemisphere. The patient had to choose the most appropriate small picture. [T]he right hemisphere—that is, the left hand—correctly picked the shovel for the snowstorm; the right hand, controlled by the left hemisphere, correctly picked the chicken to go with the bird's foot. Then we asked the patient why the left hand—or right hemisphere—was pointing to the shovel. Because only the left hemisphere retains the ability to talk, it answered. But because it could not know why the right hemisphere was doing what it was doing, it made up a story about what it could see—namely, the chicken. It said the right hemisphere chose the shovel to clean out a chicken shed.

Questions

1. In the study described above, was the snowstorm shown in the left visual field or right visual field? Why?

2. The real import of this Beyond the Numbers is contained neatly in the last two sentences of the Gazzaniga excerpt. Explain why the results Gazzaniga cites above are evidence that we are hardwired toward slippery thinking.

3. Explain why the results Gazzaniga cites above are evidence that Human Inference is potentially a far more complex task than many realize.

Why Numeracy Matters  

Name: ____   Section Number: __________

To be graded, all assignments must be completed and submitted on the original book page.

EXHIBIT 1

Nursing Knowledge Needed ________________________________

On June 2, 2007, a nurse in Wales accidentally injected an 85-year-old patient with a lethal amount of insulin during a home visit. The syringes used to inject insulin are typically marked with insulin units instead of the usual milliliters. Finding herself without an insulin syringe on hand, the nurse retrieved a new syringe from her car. She did not notice, however, that the syringe she grabbed was marked in milliliters instead of insulin units.

Questions

1. If 1 milliliter equals 100 insulin units, how many milliliters should the patient have been given if she had been prescribed 36 units? Show your work.

2. The nurse injected the patient four times with a full 0.9 milliliter syringe. What was the nurse's mistake? Defend your answer.

EXHIBIT 2

Statistical Citizenship ________________________________

Read the article "Democracy and the Numerate Citizen: Quantitative Literacy in Historical Perspective," by Professor Patricia Cline Cohen: https://www.maa.org/external_archive/QL/pgs7_20.pdf. If the link is broken, search for the article on the web or ask for help.

1. Briefly list three "features of the Constitution that suggest a numerical approach to governance."

2. Initially, the government was reluctant to collect more than the most basic census information of race, sex, and age. Why? During which of the three time periods addressed by Cohen did this attitude change?

3. Cohen writes that "the post-Civil War era finally brought a full melding of statistical data with the functioning of representative government." List three facts supporting this claim.

4. Refer to Appendix A from Cohen's article. Which of the 12 suggestions to promote quantitative literacy would have the greatest impact? Defend your answer.

Slippery Evidence and Confounding 

Name: _______________________________________   Section Number: _________

To be graded, all assignments must be completed and submitted on the original book page.

Carefully read and think about each of the exhibits below. Then, give detailed answers to the questions that accompany each exhibit.

EXHIBIT 1

Questions

1. Looking at the Results section, what inference are you likely to make about the effectiveness of online instruction?

Thinking Critically ___________________________________

Title: Learning in an Online Format versus an In-Class Format: An Experimental Study

Authors: Allan H. Schulman and Randi L. Sims

Source: *T.H.E. Journal* 26, no. 11 (1999): 54–56

Methodology Students enrolled in five different undergraduate online courses during the Fall semester 1997 participated in a voluntary test-retest study designed to measure their learning of the course material. These students were compared with students enrolled in traditional in-class courses taught by the same instructors.

Subjects In total, 40 undergraduate students were enrolled in the online courses and 59 undergraduate students were enrolled in the in-class courses during the testing period.

Pre-tests Instructors designed pre-tests to measure the level of knowledge students had of the course content prior to the start of the course. The average pre-test scores for online students was 40.70 (s.d. = 24.03). The average pre-test scores for in-class students was 27.64 (s.d. = 21.62).

Post-tests Instructors designed post-tests on a 100-point scale to test students' knowledge at the end of the course. The average post-test scores for online students was 77.80 (s.d. = 18.64). The average post-test scores for in-class students was 77.58 (s.d. = 16.93).

Results [O]ur results indicate that there were no significant differences for post-test scores. …

2. Give at least two reasons why this inference might be compromised. Be sure your reasons come from the part of this paper that you have access to here.

"Make Mine a Large" 

In the 2009 *New York Times* piece "Excess Pounds, but Not Too Many, May Lead to Longer Life," author Roni Caryn Rabin reported:

> Being overweight won't kill you—it may even help you live longer. That's the latest from a study that analyzed data on 11,326 Canadian adults, ages 25 and older, who were followed over a 12-year period. The report … found that overall, people who were overweight but not obese—defined as a body mass index of 25 to 29.9—were actually less likely to die than people of normal weight, defined as a B.M.I. of 18.5 to 24.9.
>
> By contrast, people who were underweight, with a B.M.I. under 18.5, were more likely to die than those of average weight. Their risk of dying was 73% higher than that of normal weight people.

Question

1. Although this article doesn't describe an experiment, it does imply that being a little overweight may lead to a longer life. Identify at least one confounding variable that may compromise the validity of this inference. Support your case.

Confounding Confusion 

Name: ____ Section Number: _________

To be graded, all assignments must be completed and submitted on the original book page.

Background

When it is possible to achieve, randomization is critical to experimentation. In some cases, experiments aren't really experiments at all, but are observational studies that compare two groups of data that have already been collected. In other cases, new experimental data are compared to existing data. In still other cases, randomization is simply not possible for ethical or practical reasons. In all of these situations, the potential severity of confounding must be evaluated on a case-by-case basis.

EXHIBIT 1

Brains and Beats

Do children who study music perform better in school? One of the studies supporting this claim was conducted by University of California professor Gordon Shaw and reported in a 1999 edition of the *Deseret News.* In this study, students in the 95th Street school, one of Los Angeles' 100 poorest-performing institutions, received both piano lessons and automated mathematics training. Their ability to understand and analyze ratios and fractions was then compared to a 1997 study involving students from under-achieving schools in Orange County who were given automated and traditional mathematics instruction (but no musical training). The *News* reported that "[t]he Los Angeles students scored 2% higher than their Orange County counterparts in their ability to understand and analyze rations and fractions—concepts usually not introduced until sixth grade."

Questions

1. How do you know that the subjects in this study could not have been randomized to the treatments that were compared?

2. Aside from randomization issues, list two other possible sources of confounding that might challenge the conclusion that exposure to music caused the L.A. students to do better.

Fuzzy Quasi Is a Bear

Quasi experiments are studies that are unable to use randomization to evaluate effectiveness of interventions. This can make it difficult to tease out possible confounders and assess the integrity of any cause and effect claims. Consider this example adapted from a paper in *Clinical Infectious Diseases.*[1] A hospital wants to know if providing alcohol-based hand cleaners for staff will reduce the rate at which bacterial infections occur in the patient population. The hospital needs to design a study that will address this question.

Questions

1. Suppose a researcher suggests to place these cleaners in randomly chosen patient rooms and not in others. At the end of the study, the infection rates for each group could be compared. Why might this be an impossible design to implement in practice? Give one reason and explain.

2. The hospital is more likely to adopt a quasi-design that collects data for an extended period of time before hand cleaner dispensers are installed, and then compares those findings with data collected after the dispensers are installed. State and explain two possible sources of confounding that might occur with this kind of design.

3. In the absence of random assignment, arguments addressing confounding often have to be made some other way. What are two arguments that one might be able to make to support results obtained from a quasi-design like the one in Question 2?

[1] "The Use and Interpretation of Quasi-Experimental Studies in Infectious Diseases," volume 38, Issue 1, pp. 1586–1591.

Catching on to Experimentation  

Name: ____   Section Number: _________

To be graded, all assignments must be completed and submitted on the original book page.

A Measured Response

Twelve volunteers are needed for this study. Your instructor will randomly divide the twelve persons into two groups, Group R and Group L. Group R will perform this experiment with their right hands and Group L will use their left hands.

Instructions

1. Choose a subject to begin. Make sure that he or she knows which hand to use.

2. Subject extends his or her hand and hold out the thumb and forefinger in a pinching stance (about 1" gap).

3. A group member will hold a ruler above the subject's hand so that when the ruler is let go, the subject will be able to catch it by pinching his or her fingers together. The 0 centimeter mark on the ruler should be level in the between the subject's fingers.

4. Subject will be instructed to catch the ruler with his or her two fingers as soon as possible after it is dropped. No advanced notice about release will be given.

5. Record the position of the subject's fingers on the ruler when he or she catches it. Convert that number to a reaction time in seconds using Table 1.10, and record it in Table 1.11.

6. Repeat until all twelve subjects have performed.

TABLE 1.10 Experiment Results

Distance (cm)	Time (sec)	Distance (cm)	Time (sec)
1	0.045	16	0.181
2	0.064	17	0.186
3	0.078	18	0.192
4	0.090	19	0.197
5	0.101	20	0.202
6	0.111	21	0.207
7	0.119	22	0.212
8	0.128	23	0.217
9	0.135	24	0.221
10	0.143	25	0.226
11	0.150	26	0.230
12	0.156	27	0.235
13	0.163	28	0.239
14	0.169	29	0.243
15	0.175	30	0.247

TABLE 1.11 Experiment Results

Group R	Time (sec)	Group L	Time (sec)
1		1	
2		2	
3		3	
4		4	
5		5	
6		6	

Questions

1. The data in the Distance/Time table come from the formula:

$$\text{Distance} = \frac{1}{2}at^2$$

where "t" is time, in seconds, and "a" is acceleration due to gravity (9.8 meters/sec). Verify that the time t corresponding to a distance of 4 cm is 0.090 sec. Show all your work.

2. What is the response variable in this experiment?

3. What is the explanatory variable?

4. What is a potential confounding variable? How might this have affected the experiment?

5. Find the means of both groups. Based on those two values, is there evidence of a difference between the reaction times of Group L and Group R? Defend your answer.

6. What role does the variance of the measurements in each group have in the precision of the assessment in Question 5? Explain.

7. Instead of having six people use their left hands and six use their right hands, the experiment could have been designed so that all twelve volunteers used both their right and left hands, in random order. State and defend two reasons why this might have been a better design to have used.

Questionable Evidence 

Name: ____   Section Number: __________

To be graded, all assignments must be completed and submitted on the original book page.

EXHIBIT 1

Cancer Carafe

On March 12, 1981, the *New York Times* reported on a Harvard study that linked coffee consumption and pancreatic cancer. Article author Harold Schmeck Jr. noted that "[t]he report estimated that more than half of the pancreatic cancer cases that occurred in the United States might be attributable to coffee drinking …." To complete the assignment, begin by reading Schmeck's article: http://www. nytimes.com/1981/03/12/us/study-links-coffee-use-to-pancreas-cancer.html. If this link does not work, search for the article under its title "Study Links Coffee Use to Pancreas Cancer," or by its author.

Questions

1. What two groups were being compared in this experiment?

2. List at least two other sources of circumstantial evidence that were cited in the article as further support of a link between pancreatic cancer and coffee.

3. Schmeck followed up on this article with another one titled "Critics Say Coffee Study Was Flawed." List three potential sources of confounding mentioned in this article and comment on why these could potentially destroy any claim to cause and effect? You can find the article online at: http://www.nytimes.com/1981/06/30/science/critics-say-coffee-study-was-flawed.html?module=Search &mabReward=relbias%3Ar. Use the space on the next page, if needed, to complete your answer.

Of Mice and People

Read the article "Misleading Mouse Studies Waste Medical Resources" by Erika Hayden, which appeared in *Nature* on March 26, 2014. You may find it at: http://www.nature.com/news/misleading-mouse-studies-waste-medical-resources-1.14938.

Questions

1. The article addresses amyotrophic lateral sclerosis (ALS), suggesting that mice studies might be misleading. The article gives two reasons why. List both of those here.

2. A 1958 amendment to the Food, Drugs, and Cosmetic Act of 1938 called the "Delaney Clause" has been instrumental in the banning of food additives since its enactment. Find and state the exact text of the one-sentence Delaney Clause.

3. In February 2014, sandwich giant Subway announced it would stop using azodicarbonamide in its breads. Research this decision and comment on the indirect role that the Delaney Amendment had in Subway's decision. You may be able to find CNN article by Elizabeth Landau about this issue at: http://www.cnn.com/2014/02/06/health/subway-bread-chemical/.

4. Do you believe that misleading studies can still be beneficial to society? If so, explain some potential benefits of misleading studies. If not, explain what can be done to limit the number of misleading studies that are conducted.

Assessing Statistical Significance  

Name: ____   Section Number: _________

To be graded, all assignments must be completed and submitted on the original book page.

EXHIBIT 1

Badge of Big

Statistical science endeavors to answer the question "are treatment results different enough that they are unlikely to have occurred by chance?" If so, the results are said to be statistically significant. In Beyond the Numbers 1.8 your class may have completed an activity where twelve reaction times were computed—six for a group using their left hands and six for a group using their right hands. Your instructor may want you to use those values rather than the twelve results recorded in the table below. Follow her instructions.

Questions

1. Find the mean of the six Group R measurements.

2. Find the mean of the six Group L measurements.

TABLE 1.13 Experiment Results

Group R	Time (sec)	Group L	Time (sec)
1	0.090	1	0.111
2	0.119	2	0.181
3	0.143	3	0.090
4	0.169	4	0.186
5	0.064	5	0.045
6	0.150	6	0.143

3. Find the variance of the six Group R measurements. You will probably want to use a software package. The variance is the square of the standard deviation. cathing on .

4. Find the variance of the six Group L measurements. You will probably want to use a software package. The variance is the square of the standard deviation.

5. Compute:

$$Z = \frac{(\text{mean of Group R}) - (\text{mean of Group L})}{\sqrt{\dfrac{(\text{variance of Group R})}{6} + \dfrac{(\text{variance of Group L})}{6}}}$$

6. The difference between the left-hand reaction times and right-hand reaction times is not statistically significant if $-2.23 < |Z| < 2.23$, where Z is what you computed in Question 5. Do the results shown in Table 1.13 support a statistically significant difference in reaction times? Why or why not?

7. What does statistical significance mean in the context of his activity?

8. Go online to research a study that found (or did not find) statistical significance. Describe that study and explain the implications of the finding.

Instructor's Note: The notation Z was used instead of "t" to avoid confusion later in the workbook. Also the degrees of freedom were computed using the simple estimate of $6 + 6 - 2$.

Designer Thoughts  

Name: ____   Section Number: __________

To be graded, all assignments must be completed and submitted on the original book page.

Background

Whether an experiment is found to produce statistically significant results is not just a function of the effectiveness of the treatments being evaluated. The design of the experiment and how the data are analyzed can make a huge difference as well. The data on the right represent the reaction times of subjects playing an online game designed to test hand-eye coordination.

Pairing Profits

There are two ways to look at these data.

Scenario A: As 24 individuals randomly assigned to two groups (left and right hand usage)

Scenario B: As 12 individuals who each had to use both their left and right hands in a paired experiment

Let's see why it matters. The decompositions below determine whether the results are statistically significant, or not. Some technical details have been compromised so that the presentation remains accessible.

If which hand is being used is going to show up as statistically significant in this experiment, then the ratio of the "Variance Attributed to the Hand" to the "Variance Left Unexplained" is going to have to be 0.45 or larger.[1]

TABLE 1.14 Reaction Times

Left	Right
1	1.05
0.74	0.76
0.66	0.71
0.78	0.79
0.68	0.69
0.65	0.72
0.75	0.75
0.69	0.72
0.94	0.99
0.79	0.8
0.81	0.82
0.62	0.67

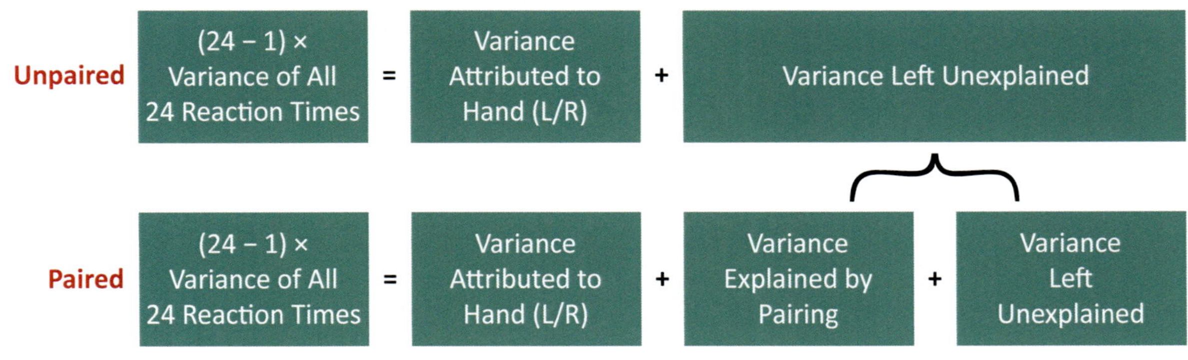

[1] Instructor's Note: "Degrees of freedom" are conservatively taken to 1 and 11 so that a discussion of degrees of freedom can be avoided.

Questions

1. Verify that the variance of all 24 observations is 0.013234. Use a software package such as Microsoft Excel or Apple Numbers and explain your work.

2. Let's look at the data as in Scenario A. If the Variance Attributed to Hand is 0.005400, what is the ratio of "Variance Attributed to Hand" to the "Variance Left Unexplained?" Is this ratio big enough to say there is a statistically significant difference in reaction times?

3. Let's look at the data as in Scenario B. If the Variance Attributed to Hand is 0.005400 and the variance explained by the pairing is 0.29608, what is the ratio of "Variance Attributed to Hand" to the "Variance Left Unexplained?" Is this ratio big enough to say there is a statistically significant difference in reaction times?

4. Take a few moments to reflect on what you've just shown. What are the implications for human inference if paired data were not analyzed as paired (or an experiment wasn't designed as paired when it could have been)?

What to Believe? 

Name: ____   Section Number: __________

To be graded, all assignments must be completed and submitted on the original book page.

Background

When it comes to experimentation, principled implementation and ethical reporting are obviously critical to the integrity of all associated human inference. Unfortunately, there are a surprisingly large number of cases where dishonesty or exploitation, not just carelessness or confounding, tainted or nullified the significance of the findings. You will investigate three examples of research misconduct in this activity.

EXHIBIT 1

Piltdown Meltdown—1912

In 1912, Charles Dawson discovered two skulls found in the Piltdown quarry in Sussex, England. These skulls were said to be from a primitive hominid and were hailed as the "missing link" between man and ape. Research the Piltdown Man and write a well-formed summary of what ultimately was revealed about the skulls. Give specific details about the nature of the deception that was uncovered. How does this case affect your understanding of statistics going forward?

Marker Mice—1974 ___________

William T. Summerlin used to work at Memorial Sloan Kettering Cancer Center in New York City, conducting research in transplantation immunology. His work could have had major implications for reducing the rejection rates of transplanted tissue. Research Dr. Summerlin and write a well-formed summary of what ultimately was revealed about his work. Give specific details about the nature of the deception that was uncovered. What caused Summerlin to behave as he did, and what can be done to avoid such deceptions?

Doing the Dishes—2010 ___________

Dr. Vipul Bhrigu was a researcher at the University of Michigan when he began to feel professionally threatened by the work of a graduate student in the same lab, Heather Ames. Desperate for his work not to be overshadowed, he concocted a plan to keep hers from moving ahead. Research Dr. Bhrigu and write a well-formed summary of what ultimately was revealed about his deception. Give specific details about the nature of the sabotage that was uncovered. How can this type of sabotage be avoided?

Background Bugaboos

Name: ____   Section Number: __________

To be graded, all assignments must be completed and submitted on the original book page.

Violence in Chicago

Title: Number of CPS Students Shot Rises, as Does Fear of More to Come

Authors: Noreen S. Ahmed-Ullah

Source: *Chicago Tribune,* June 26, 2012, http://articles.
chicagotribune.com/2012-06-26/news/ct-met-cps-student-
violence-0625-20120626_1_cps-students-students-shot-safe-
haven-program

Twenty-four students were fatally shot during the school year that ended
June 15, four fewer than in the 2010–11 year. But the overall shooting
toll—319—was the highest in four years and a nearly 22% increase from
the previous school year.

Questions

1. The "previous school year" being referred to is the 2010–11 year. What was the total number
   of shootings during the 2010–11 school year? Make sure you explain how you got your answer.

2. Speaking of violence in Chicago, the table below shows the number of murders in Chicago from 2001 through 2012. Find the mean and the median number of murders over this time period. Why are these two numbers so different for these data? Look up these terms if you don't remember what they mean.

TABLE 1.1 Violence in Chicago

Year	2001	2002	2003	2004	2005	2006	2007	2008	2009	2010	2011	2012
Murders	667	656	601	453	451	471	448	513	459	436	435	506

3. Suppose you wanted to use a chart to summarize the above data about murders in Chicago. Would a bar chart or a histogram be more appropriate? Defend your answer, and use the space below to construct the chart you've chosen. If you don't remember the difference between the two types of charts, look it up.

Association and Causation 

Name: ____   Section Number: __________

To be graded, all assignments must be completed and submitted on the original book page.

Carefully read and think about each of the exhibits below. Then, give detailed answers to the questions that appear at the very end.

EXHIBIT 1

Breakfast and Biology

Title: Does Eating Breakfast Affect the Performance of College Students on Biology Exams?

Author: Gregory W. Phillips, Blinn College

Source: *Bioscience* 30, no. 4 (2005): 15–19

Abstract This study examined the breakfast eating habits of 1,259 college students over an eleven-year period to determine if eating breakfast had an impact upon their grade on a General Biology exam.

…

Results and Discussion This study showed that students who ate breakfast had a higher success rate on General Biology exams than those students who did not eat breakfast. This finding supports earlier research, which indicated that eating breakfast affects student performance.

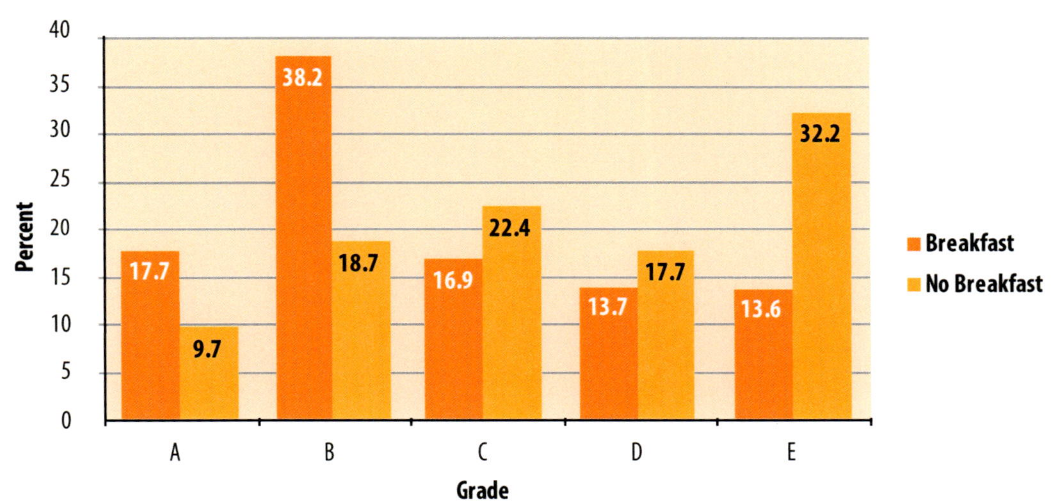

Source: Gregory W. Phillips "Does Eating Breakfast Affect the Performance of College Students on Biology Exams?" *Bioscience* 30, no. 4 (2005): 15–19.

1. Based solely on the graph, could you say eating breakfast *affects* student performance? Explain.

EXHIBIT 2

Mortality and Global Warming

On November 14, 2007, the Kentucky legislature held hearings on global warming with speakers Christopher Walker Monckton and James Taylor of the Heartland Institute.

Look at the graph below, allegedly typical of the exhibits that Monckton and Taylor showed.

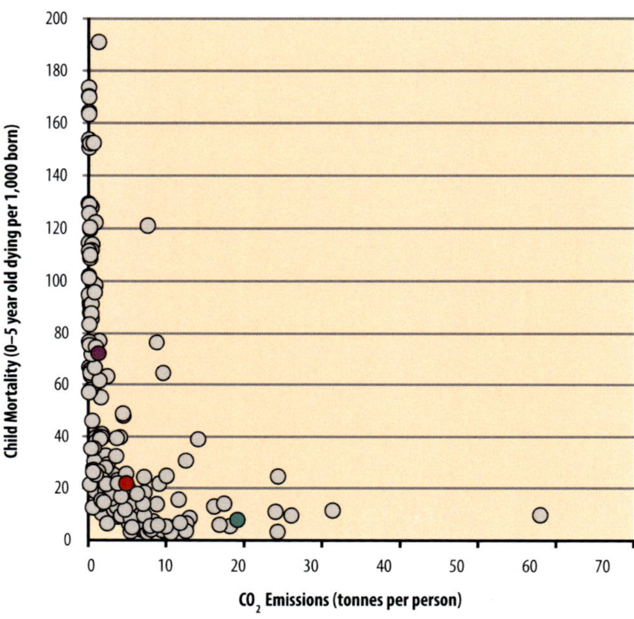

For example, India is the blue circle centered at 1.3 tonnes of CO_2 per person (*x*-value) and 72 child/infant deaths per 1,000 births (*y*-value). You should take a close look at the trend exhibited by the plot. The red circle represents China and the green circle represents the United States.

Question

1. Reportedly, the Legislature was told that this graph shows that global warming is good because it *causes* a decrease in infant mortality. Is this an accurate assessment? Explain.

Confidence Intervals

2

Overarching Goal

The primary intent of this module is to develop a broad understanding of what statistical confidence means, what it doesn't mean, and what components are required for its construction.

Learning Outcomes

You'll know you have successfully completed this module when you are able to:

1. **Define** and **demonstrate** simple random sampling.

2. **Identify** and **analyze** alternative sampling methods.

3. **Explain** the difference between randomness and representativeness.

4. **Define** sampling variability, and **explain** the role it plays in the construction of a confidence interval.

5. **Define** sampling distribution, and **explain** the role it plays in the construction of the margin of error.

6. **Compute** and **interpret** confidence intervals for a proportion or mean.

7. **Define** and **apply** the empirical rule to solve probability problems.

8. **Identify** categorically good or bad surveys, and **explain** the reasons they are so categorized.

9. **Explain** the difference between sampling variability and non-sampling variability.

10. **Identify** and **evaluate** strategies for addressing non-sampling variability.

Get MOE Out of the Way  

Name: ____   Section Number: __________

To be graded, all assignments must be completed and submitted on the original book page.

EXHIBIT 1

Texting Error

Title:	Poll Finds Support for Ban on Texting at the Wheel
Author:	Marjorie Connelly
Source:	*New York Times*, September 27, 2009, http://www.nytimes.com/2009/09/28/technology/28truckerside.html

Read the following extract from the above article and answer the related questions to see if you understand the data.

> The public overwhelmingly supports the prohibition of text messaging while driving, the latest *New York Times*/CBS News Poll finds. Ninety percent of adults say sending a text message while driving should be illegal, and only 8% disagree.
>
> …
>
> The *Times*/CBS News telephone poll was conducted September 19–23 with 1,042 adults nationwide and has a margin of sampling error of plus or minus three percentage points.

Questions

1. How was the sample taken and what was the result of the survey?

2. Suppose someone said to you, "Sure, of the 1,042 surveyed by the poll, 90% agreed, but I bet if you interviewed all American adults you would likely find only 50% agreeing!" Is this an accurate assessment and why?

EXHIBIT 2

A Weak Majority

Title: Poll Finds Slim Majority Back More Afghanistan Troops

Author: Adam Nagourney and Dalia Sussman

Source: *New York Times*, December 9, 2009, http://www.nytimes.com/2009/12/10/world/asia/10poll.html

Read the following extract from the above article and answer the related questions.

> A bare majority of Americans support President Obama's plan to send 30,000 more troops to Afghanistan, but many are skeptical that the United States can count on Afghanistan as a partner in the fight or that the escalation would reduce the chances of a domestic terrorist attack, according to the latest *New York Times*/CBS News poll.
>
> …
>
> The support for Mr. Obama's Afghanistan policy is decidedly ambivalent, and the nation's appetite for any intervention is limited. Over all, Americans support sending the troops in by 51% to 43%, while 55% said setting a date to begin troop withdrawals was a bad idea.
>
> …
>
> The poll was conducted by telephone from Friday through Tuesday night, with 1,031 respondents, and has a margin of sampling error of plus or minus three percentage points.

Questions

1. About how many people in the sample supported sending more troops to Afghanistan?

2. Suppose the headline had read as follows: "Poll Finds Slim Majority of All Americans Back More Afghanistan Troops." Offer some objections to that.

Are Online Reviews Statistical Samples? 

Name: ____   Section Number: _________

To be graded, all assignments must be completed and submitted on the original book page.

EXHIBIT 1

Bravos for Bucks ___________________________

The VIP brand Kindle Fire cover received 4,945 reviews on Amazon by early 2012, averaging a nearly perfect 4.9 stars out of five. That's quite impressive. It is tempting to think that online reviews, especially those posted at major sites like Amazon, are representative of consumer experiences. We know, however, that voluntary responses are often biased. But are product reviews even less accurate than previously though? In his 2012 *Time* article "9 Reasons Why You Shouldn't Trust Online Reviews," Brad Tuttle writes, "You shouldn't believe everything you read. And if you're reading online reviews of products, hotels, restaurants, or local businesses or services? Then you should believe even less." You can find Tuttle's article online at: http://business.time.com/2012/02/03/9-reasons-why-you-shouldnt-trust-online-reviews/.

Questions

1. Describe three reasons listed in the article as to why you should be very cautious about online reviews.

2. What was VIP doing to boost the ratings of its Kindle Fire cover? Be specific.

3. Compared to computer algorithms, how well did people perform in spotting fake reviews? How does this study inform your perception of online reviews?

Bosom Bezos Buddy

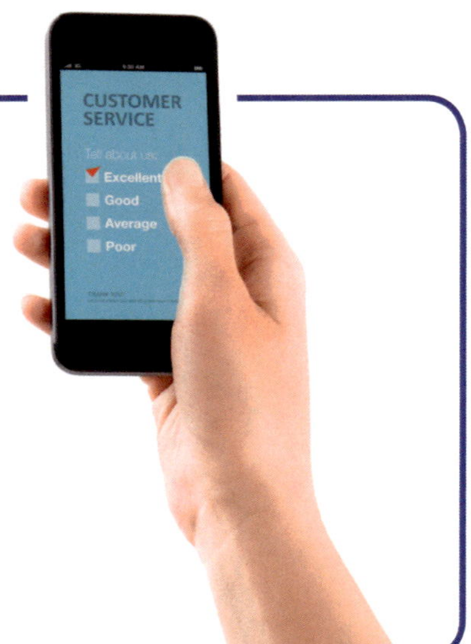

Fixing the problem of biased online product reviews is not going to be easy to do. Let's pretend for a moment that the founder and CEO of Amazon, Jeff Bezos, hired you to devise a solution that captures what you know about statistical sampling. Offer Mr. Bezos a sampling plan that is statistically defensible. Make sure you identify your population, your samples, and how you plan to select them. Identify, too, the parameter and statistic(s) of interest. Explain why your sampling plan is better than what is currently being done, and describe how it will correct for some of the problems you saw in Exhibit 1.

Random or Representative? 

Name: _______________________________________   Section Number: __________

To be graded, all assignments must be completed and submitted on the original book page.

Gulliver Travels

900 people live in Gulliver, a small town in Michigan's Upper Peninsula.[1] You want to know what proportion of Gulliver's population supports legalizing marijuana. Suppose you already know the following demographic information about Gulliver's 900 citizens:

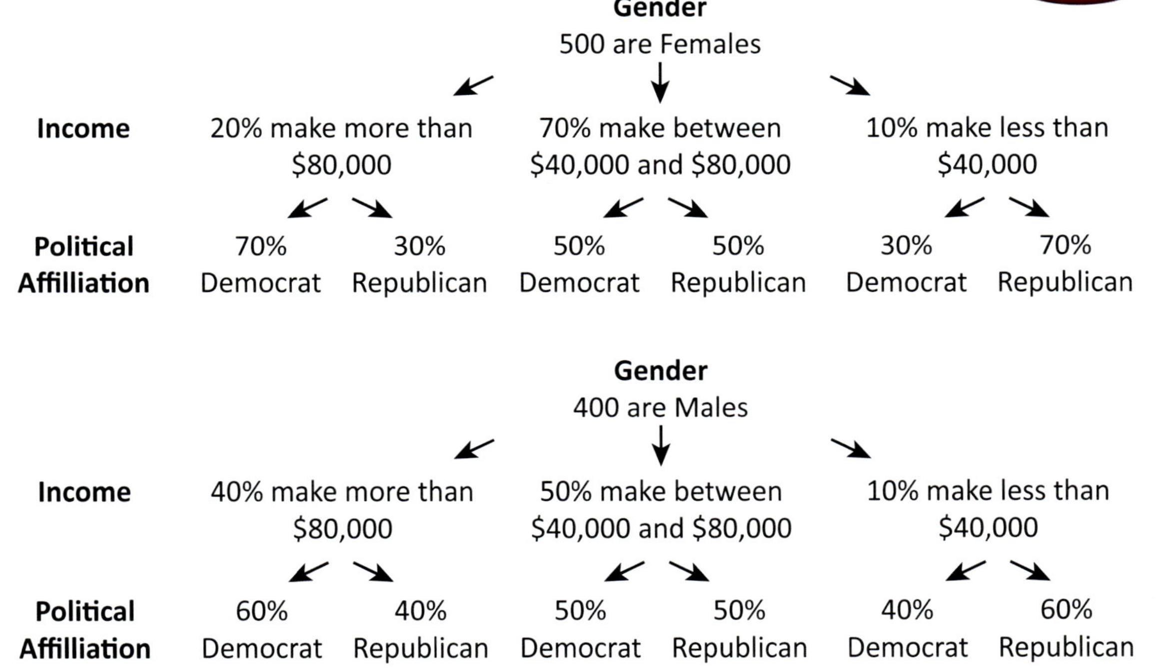

Questions

1. You have enough money to interview 90 residents. Working much the way Gallup did in the 1930s you want your sample of 90 to mirror the distribution of subjects in the population exactly (at least along the lines of gender, income, and political affiliation). How many people would your sample place in the groups shown on the next page? If a calculation results in a partial person (e.g., 6.4 persons), leave the number as it is—don't round.

---

[1] There really is a Gulliver, MI of about this size. The demographics are completely made up, however.

Category	Number of Persons
Males	
Females	
Males making between $40,000 and $80,000 yearly	
Females making less than $40,000 per year who are Democrats	
Male Republicans making over $80,000 per year	

2. Suppose the cross-sectional sample taken above represents a perfect microcosm of the larger population with respect to the legalization of marijuana. Is there any uncertainty involved in using this sample to represent the proportion of people in Gulliver who favor the legalization of marijuana? Explain.

3. Suppose you decided, instead, to take a simple random sample of Gulliver's population. Explain how you could take an SRS of size 90 from this population.

4. A carefully chosen simple random sample may not be representative of the population. Explain how this could be.

5. The legalization of marijuana is a hot-button issue for many across the country. How do you think the results of this poll could be used to better understand the debate over marijuana legalization?

Purposive Sampling 

Name: ____   Section Number: __________

To be graded, all assignments must be completed and submitted on the original book page.

EXHIBIT 1

Probability in Peril

A simple random sample is a type of *probability* sample. The upside of using probability samples is that they allow mathematical assessments of the accuracy of sample-based parameter estimates. Purposive sampling is another type of sampling. It is not probabilistic, meaning that it does not adhere to probability theory. As a result, numerical assessments of sampling integrity are typically not possible. Still, one often sees these types of samples used—particularly in social science studies—so it is important to know what they are. You will look at four types of purposive sampling in this activity:

- Convenience Sampling

- Snowball Sampling

- Heterogeneity Sampling

- Expert Sampling

Use your research skills to look up each of these types of samples and answer the questions below.

Questions

1. Define what is meant by a *convenience sample* and give an example.

2. Define what is meant by a *snowball sample* and give an example.

3. Define what is meant by a *heterogeneity sample* and give an example.

4. Define what is meant by an *expert sample* and give an example.

5. Give some well-stated reasons why you think it would be difficult to infer from these types of samples to a larger population.

Decisions with Confidence Intervals

Name: ____   Section Number: __________

To be graded, all assignments must be completed and submitted on the original book page.

EXHIBIT 1

Sticker Shock

Title: Great Jobs, Great Lives: The 2014 Gallup-Purdue Index Report

Authors: Gallup Organization and Purdue University

Source: http://products.gallup.com/168857/gallup-purdue-index-inauguralnational-report.aspx

Does it really matter where you go to college? Not so much, according to this Gallup-Purdue study. Five areas of well-being were measured among recent college graduates: purpose well-being, social well-being, financial well-being, community well-being, and physical well-being. The study concluded "that the type of schools these college graduates attended—public or private, small or large, very selective or less selective—hardly matters at all. … Just as many graduates of public as not-for-profit private institutions are thriving—which Gallup defines as strong, consistent, and progressing—in all areas of their well-being." Percentages of thriving graduates among each institution type are shown in the bar chart below.

These results were obtained from internet surveys conducted between February 4 and March 7, 2014. The margin of error is estimated to be 1%.

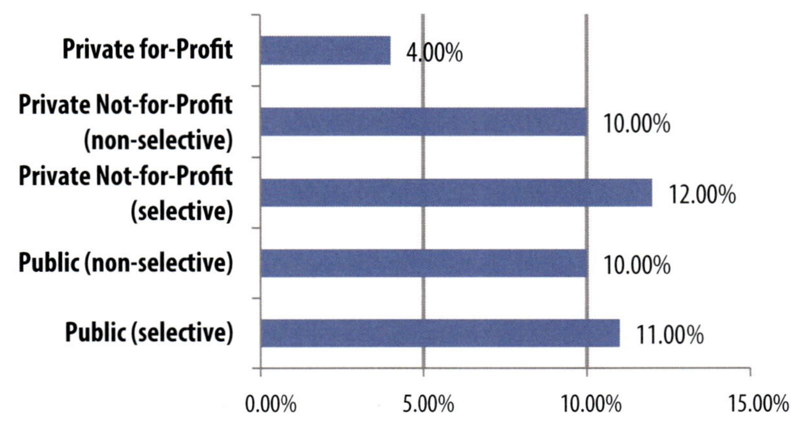

1. Compute a 95% confidence interval for the true thriving rates among graduate of public selective institutions and a 95% confidence interval for the true thriving rates among graduates of private not-for-profit selective institutions. Do these intervals overlap? Do you think this tells you anything about whether thriving rates really differ for private not-for-profit institutions and public selective institutions? Explain.

2. Compute a 95% confidence interval for the true thriving rates among graduate of public non-selective institutions and a 95% confidence interval for the true thriving rates among graduates of private for-profit selective institutions. Do these intervals overlap? Do you think this tells you anything about whether thriving rates really differ for private for-profit institutions and public institutions? Explain.

3. How does this study affect your opinion of the college application process? If you could go back and do it all over again, would you?

4. How does this study affect your opinion of the college application process? If you could go back and do it all over again, would you?

Mathematically Organic Bells: 

A Hands-On Approach

Name: ____   Section Number: _________

To be graded, all assignments must be completed and submitted on the original book page.

Normal Development

Generating the Data

You instructor may conduct this experiment in a manner different from how it is written. He may use dice, spinners, Legos, magnetic strips, or objects completely other than those described below. The point of the exercise will be unchanged. It is a good exercise to do with a partner or in small groups, provided the actual number of samples generated is sufficiently large. Your instructor will clarify how he wants this done.

1. Roll your die 20 times and record the number of times an even face appears.

2. Using the conversion chart (Table 2.6) below, convert your results to a percentage and record that number here.

Percent of Even Faces: _________________________________

TABLE 2.6

Number of Even Faces	Percentage of Even Faces	Number of Even Faces	Percentage of Even Faces
0	0	11	55
1	5	12	60
2	10	13	65
3	15	14	70
4	20	15	75
5	25	16	80
6	30	17	85
7	35	18	90
8	40	19	95
9	45	20	100
10	50		

A virtual version of this same exercise can be constructed using a spinner applet (e.g., http://www.shodor.org/interactivate/activities/BasicSpinner) and a histogram applet (e.g., http://www.shodor.org/interactivate/activities/Histogram/). Be aware that JAVA is now blocking unsigned applets. See https://www.java.com/en/download/help/java_blocked.xml for details and a confirmed workaround.

Displaying Your Result

Once you have calculated your percentage of even faces, come to the front of the room and place a Lego block in the category that corresponds to your sample percentage. (Your instructor may not use Lego blocks, as mentioned, but will demonstrate accordingly so you will know what to do.)

Questions

1. If you were to roll this die a really large number of times, about what percentage of even faces should you get?

2. What sort of real-world polling exercise could be modeled by this activity? Be sure to identify the population, sample, and statistic in that real polling context.

3. After the Lego structure is built, have a good look at it. What is its shape and where is the peak, more or less?

    Shape: _______________________________________

    Peak Is Above: _______________________________________

Confidence in Repetition:
A Hands-On Approach

Name: _______________________________________   Section Number: _________

To be graded, all assignments must be completed and submitted on the original book page.

Strip Searching for Meaning

Your instructor may facilitate this exercise in different ways. He may use dice, spinners, popsicle sticks, magnetic strips, or objects completely other than those described below. The point of the exercise will be unchanged. This is a good exercise to do with a partner or in small groups, provided the actual number of samples generated is sufficiently large. Your instructor will clarify how he wants this done in your class.

Generating the Data

To begin the exercise, please do the following:

1. Pair up with someone next to you. Each pair will receive one die. One member of your pair will roll the die 40 times while the other member records the number of times a 1, 2, 3, or 4 lands face up.

2. Switch roles, roll the die 40 more times, and record the outcomes. (It takes only about 3–5 minutes to roll a die a total of 80 times.)

3. Compute the proportion of the total 80 rolls that produced a 1, 2, 3, or 4. Keep good counts.

4. Record your sample proportion (out of 80 rolls) that came up 1, 2, 3, or 4: _____________

Displaying the Results

1. One person from each pair should come up front and take a long magnetic strip out of the collection of magnetic strips. Think of the center of this strip as the sample proportion you calculated. The strip represents a 95% confidence interval that is built around that sample proportion.

2. The other person from each pair should come up front and take a short magnetic strip out of the collection of magnetic strips. Think of the center of this strip as sample proportion you calculated. The strip represents an 80% confidence interval that is built around that sample proportion.

3. On the display board that your instructor has created, put the center of your magnetic strips, oriented horizontally, on the line that corresponds to the values of the sample proportions that you found.

A virtual version of this same exercise can be constructed using a spinner an applet (e.g., http://www.rossmanchance.com/applets/ConfSim.html). Be aware that JAVA is now blocking unsigned applets. See https://www.java.com/en/download/help/java_blocked.xml for details and a confirmed workaround.

(continued)

4. Keep the strip level with the floor and keep in mind that a lot of these strips may have to go on the display. Your instructor will demonstrate.

5. Record whether your strip overlaps the vertical line at ⅔: Yes _______ No _______. Your instructor will ask you to record this information for the class to see as well, perhaps by way of tick marks in a table or on a chart.

After everyone is finished, have a look at the class-wide display. It may be on the chalkboard or on a separate bulletin board.

Questions

1. What is the parameter and the population? Explain your answer.

2. About what percentage of the long strips overlap a vertical line through the parameter?

3. About what percentage of the short strips overlap a vertical line through the parameter?

4. Carefully articulate how "confident" you can be that a 95% confidence interval (one of the long strips) will contain the parameter?

A Challenging Interpretation

REFLECT

Name: _______________________________________   Section Number: __________

To be graded, all assignments must be completed and submitted on the original book page.

EXHIBIT 1

A Common Problem

A team of psychology researchers[1] was interested in potential misinterpretations of the term "confidence" in a confidence interval. They collected data from 442 undergraduate students, 34 graduate students, and 118 of their research-active colleagues. All subjects were presented with a "fictitious scenario of a professor who conducts an experiment and reports a 95% CI for the [population proportion] that ranges from 0.1 to 0.4. Neither the topic of study nor the underlying statistical model used to compute the CI was specified in the survey." The subjects were then asked to specify whether they agreed or disagreed with each of the following statements as interpretations of that confidence interval (CI). Here is what the investigators found.

TABLE 2.7 Percentage of Subjects Agreeing with the Statement*

Statement	First Year Students (*n* = 442)	Master Students (*n* = 34)	Researchers (*n* = 118)
1. The probability that the true proportion is greater than 0 is at least 95%.	51%	32%	38%
2. The probability that the true proportion equals 0 is smaller than 5%.	55%	44%	47%
3. There is a 95% probability that the true proportion lies between 0.1 and 0.4.	58%	50%	59%
4. We can be 95% confident that the true proportion lies between 0.1 and 0.4.	49%	50%	55%
5. If we were to repeat the experiment over and over, then 95% of the time the true porportion falls between 0.1 and 0.4.	66%	79%	58%

[1] Hoekstra, R., Morey, R., Rouder, J., and Wagenmakers, E-J. "Robust misinterpretation of confidence intervals," *Psychon Bull Rev.* Published online January 14, 2014. http://www.ejwagenmakers.com/inpress/HoekstraEtAlPBR.pdf. Table has been reformatted, one statement omitted, and "mean" changed to "proportion" throughout.

All of the statements are wrong! Students and professional researchers alike found the interpretation of a confidence interval to be challenging. Yet, it is not acceptable to step away from this challenge. Confidence intervals are used everywhere as a kind of statistical seal of approval for survey and experiment results.

1. What is wrong with Statement 1? Be very clear. Look back at your content on confidence intervals to make sure that you have the right interpretation.

2. What is wrong with Statement 2? Be very clear. Look back at your content on confidence intervals to make sure that you have the right interpretation.

3. What is wrong with Statement 3? Be very clear. Look back at your content on confidence intervals to make sure that you have the right interpretation.

4. What is wrong with Statement 4? Be very clear. Look back at your content on confidence intervals to make sure that you have the right interpretation.

5. What is wrong with Statement 5? Be very clear. Look back at your content on confidence intervals to make sure that you have the right interpretation.

The Empirical Rule  

Name: ____   Section Number: ___________

To be graded, all assignments must be completed and submitted on the original book page.

Background

A bell-shaped distribution is characterized by where it peaks (mean) and how spread out it is (standard deviation). We already know that a bell-shaped sampling distribution is important to the construction of a margin of error and the associated confidence interval. However, bell-shaped distributions also contain useful probabilistic information about the variable being described. The following well-known rule addresses this connection.

Empirical Rule:

Suppose a bell-shaped distribution has a mean μ and a standard deviation σ. Then:

a. About 68% of all observations represented by that distribution will fall within one standard deviation of the mean.

b. About 95% will fall within two standard deviations of the mean.

c. About 99.7% will fall within three standard deviations of the mean.

Graphically:

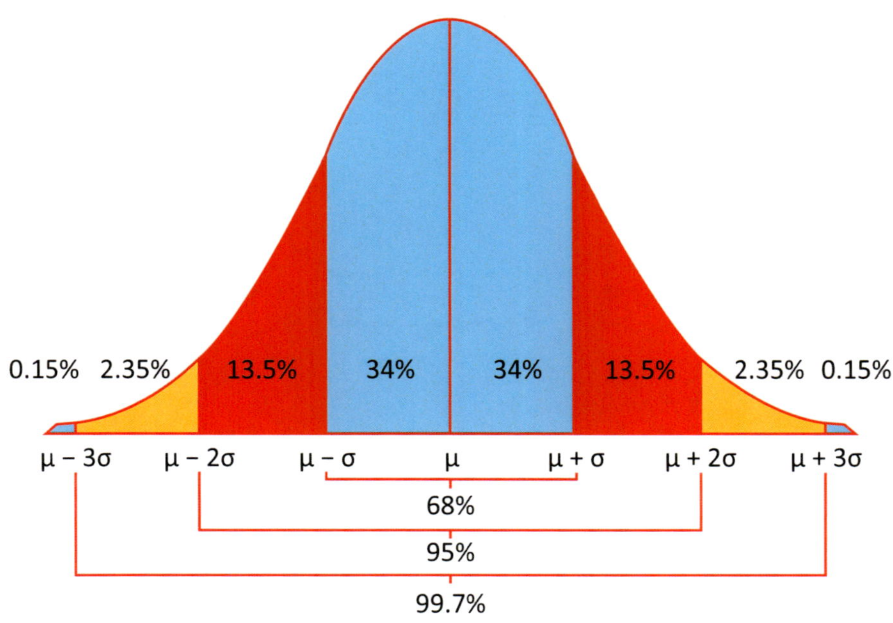

Face in Class Book ___________

In a 2012 *Washington Post* article entitled "Is College Too Easy? As Study Time Falls, Debate Rises," Daniel de Vise reports that "over the past half-century, the [average] amount of time college students actually study—read, write and otherwise prepare for class—has dwindled from 24 hours a week to about 15 ...." No standard deviation is given, but let's assume that standard deviation is 2.5 hours.

Questions

1. Suppose a college student is selected at random. Use the empirical rule to estimate how likely it is that this student studies between 10 and 17.5 hours per week.

2. Suppose a college student is selected at random. Use the empirical rule to estimate how likely it is that this student studies between 17.5 and 20 hours per week.

3. Suppose a college student is selected at random. Use the empirical rule to estimate how likely it is that this student studies more than 20 hours per week.

4. Estimate the average number of hours you study each week. How many standard deviations away from the mean do you fall?

Class on Facebook

A 2012[1] study measured "the efficacy of social networking systems as instructional tools." The study surveyed 186 students about the use of social networking systems as an active part of the semester class structure. One question asked and answered by 181 of the 186 students, along with the results received, is shown below.

Question from the study: There are no specific benefits that make Facebook a better forum for class discussions and announcements than a learning management system like Blackboard. Do you agree or disagree?

TABLE 2.8 Survey Results

Response	Number of Subjects Choosing This Response
1 – Strongly Disagree	9
2 – Disagree	43
3 – Neutral/Undecided	59
4 – Agree	52
5 – Strongly Agree	18

Make sure you can read the table. If you are having trouble interpreting the data, ask your instructor to explain it. The mean of these 181 answers is 3.15 and the standard deviation is 1.05.

Questions

1. Use the empirical rule to estimate how likely it is that an answer to this question will be in the interval 2.10 to 4.20. What was the actual percentage of answers in this interval?

2. Use the empirical rule to estimate how likely it is that an answer to this question will be above 4.20. What was the actual percentage of answers in this interval?

[1] Buzzetto-More, N. "Social Networking in Undergraduate Education," *Interdisciplinary Journal of Information, Knowledge, and Management* Volume 7, 2012.

3. This example is somewhat atypical since only five outcomes are possible and the empirical rule doesn't strictly apply. It still provides useful estimates, however. Graph the distribution of these 181 answers and show that it is, indeed, bell-shaped. Use computer software (such as Microsoft Excel or Apple Numbers) to create this plot. Cut and paste the plot (literally or electronically, depending on how you are instructed to turn this exercise in the space below.

4. Confirm that the mean is 3.15 and the standard deviation is 1.05, as claimed. Use computer software (such as Microsoft Excel or Apple Numbers) to do this. Provide detailed instructions as to how you accomplished this task.

Do Incentives Help? 

Name: ____ Section Number: _________

To be graded, all assignments must be completed and submitted on the original book page.

EXHIBIT 1

Pay to Play Early

Title: Efficacy of Incentives in Increasing Response Rates

Authors: Fahimi, M., Whitmore, R.W., Chromy, J.R., Siegel, P.H., & Cahalan, M.J. (2006).

Source: Presented at Second International Conference on Telephone Survey Methodology, Miami, FL, http://www. rti.org/publications/abstract.cfm?pubid=6019

An internet search on the effects of incentives on increasing response rates will reveal a lot of research on the topic. One particular study conducted at the Research Triangle Institute in Raleigh-Durham, North Carolina is particularly interesting. In Phase I of the study 1,197 subjects were contacted and asked to complete a survey. However, those subjects were randomly split into three subgroups. Subjects in the first group were not offered any monetary reward for early completion of the survey. Subjects in the second were offered $20, for early completion, and those in the third group were offered $30 for early completion. Here is what the researchers found with respect to the number of subjects who responded to the survey:

TABLE 2.10 Research Results

Incentive Group (Early Response)	Number of Respondents	Number of Non-Respondents
Group 1 ($0)	66	336
Group 2 ($20)	120	271
Group 3 ($30)	138	266
Total	324	873

Questions

1. Use data from the table to support or refute a claim that incentives matter for early completion of a survey.

2. Use data from the table to support or refute a claim that a high incentive is more effective than a low incentive at achieving early completion of a survey.

No Pay Replay

There were 873 non-respondents to the survey described above. In Phase II of the study, these 873 subjects were contacted a second time and asked to complete the survey. No additional incentives were offered, and the offers of incentives for early completion had already expired. Here is what the researchers found:

TABLE 2.11 Research Results

Incentive Group (Early Response)	Number of Respondents	Number of Non-Respondents
Group 1 ($0)	109	227
Group 2 ($20)	91	180
Group 3 ($30)	96	170
Total	296	577

Questions

1. Use data from the table to support or refute a claim that no-incentive follow-up requests are effective at increasing response rates.

2. Use data from the table to support or refute a claim that the effectiveness of no-incentive follow-up requests depends both on whether an incentive for early completion had originally been offered and whether that incentive was large or small.

Some Pay Saves Day

There were 577 non-respondents still remaining after the Phase II attempt. In Phase III, these 577 were contacted again and asked to complete the survey. This time, however, the remaining non-respondents were randomly divided into two groups: one would receive no compensation for completion, while the other would receive $30 for completion. Here is what the researchers found:

TABLE 2.12 Research Results

Incentive Group (Early Response)	Number of Respondents	Number of Non-Respondents
NF1 ($0)	98	190
NF2 ($30)	135	154
Total	233	344

Questions

1. Use data from the table to support or refute a claim that incentives matter when conducting follow-ups for survey completion.

2. At the end of Phase III, 233 people responded. All of these 233 had been contacted two other times over the course of the experiment. Explain why this could create confounding if complex statistical methods are not used to mitigate.

3. Take a step back and look at the results in all three of these Exhibits. Use what you have learned to describe how you would incentivize (or not) a survey that you wanted to administer. Your answer should include comments about incentives for early completion as well as incentives for follow-up completion. Remember, you are not being asked to design an experiment. That's what the authors above did. Rather, you are being asked to use what you have learned from their study to decide whether you want to attach an incentive plan to your survey, and if so, what it should look like.

Formal Inference

3
MODULE

Overarching Goal

The primary intent of this module is to understand the conceptual tenets and practical consumption of statistical hypothesis testing, beginning with the more accessible concepts of sensitivity and specificity.

Learning Outcomes

You'll know you have successfully completed this module when you are able to:

1. **Define** and **compute** sensitivity and specificity.

2. **Explain** the effect on sensitivity and specificity of changes to the testing criteria.

3. **Identify** and **demonstrate** the difference between probabilities of conditional and unconditional events.

4. **Define** Type I error and **explain** how to view hypothesis testing as a screening test.

5. **Explain** the difference between a Type I error and a p-*value.*

6. **Define** the meaning of the phrase *statistical significance.*

7. **Analyze** the use of the phrase *statistically significant* in media reports.

8. **Explain** the difference between *statistical significance* and *practical significance.*

9. **Execute** the steps needed to test simple hypotheses.

10. **Compute** and **demonstrate** the use of *p*-value when testing a hypothesis.

Bayes' Rule 

Name: _______________________________________   Section Number: _________

To be graded, all assignments must be completed and submitted on the original book page.

Background

The notation P(A|B) is read as "the probability of A, given B, has occurred." So the "|" symbol is read as "given." Formally, A and B are called *events* and P(A|B) is a *conditional* probability.

Bayes' rule is a very useful way of relating conditional and unconditional probabilities. According to this rule, for any two events A and B, we have:

$$P(A|B) = \frac{P(B|A) \times P(A)}{P(B)}$$

Let's use "T+" to denote the event "the screening test concludes that the condition (disease, pregnancy, etc.) is present." Likewise "T–" is notation for the event "the screening test concludes that the condition is not present." "CP" denotes that a condition is actually present, while "CA" denotes that a condition is actually absent. You already know some of the following definitions, though the notation is probably new to you. Others are likely all new:

1. P(T+|CP) is the **sensitivity** of the test.

2. P(T–|CA) is the **specificity** of the test.

3. P(CP|T+) is the **positive predictive value** of the test.

4. P(CA|T–) is the **negative predictive value** of the test.

5. P(CP) is the **prevalence** of the condition in the population.

EXHIBIT 1

Ottawa Ankle Rules

Title: Sensitivity of the Ottawa Rules

Authors: G. Lucchesi, R., Jackson RE, W. Peacock WF, C. Cerasani, and R. Swor

Source: *Ann Emerg Med.* 1995 Jul;26(1):1–5.

The Ottawa ankle rules, commonly used in medicine, are designed to exclude fractures of the ankle (or midfoot). The rules first require a careful examination of two particular places on the ankle. If there is tenderness in either of those two places, or an inability to bear weight on the foot, then the Ottawa rules conclude that there is a fracture. Else, the rules conclude that there is not a fracture. In this publication, the authors conducted one of the first sensitivity and specificity studies on the Ottawa ankle rules. The following table did not appear in the referenced article, but was reconstructed from summary values that were recorded therein. A total of 421 patients with ankle injuries were studied. The truth regarding their fractures was confirmed with x-rays after the Ottawa ankle rules were applied.

1. Let CP represent the event that the ankle was truly fractured, and let T+ represent the event that the test said it was. From the table, estimate the positive predictive value and the negative predictive value.

TABLE 3.10 Ankle Test Results

Predicted by the Ottawa Ankle Test	Truth Regarding Fracture		Totals
	Not Fractured	Fractured	
Not Fractured	51	5	56
Fractured	277	88	365
Totals	328	93	421

2. Which do you think a patient would be more interested in, $P(T+|CP)$ or $P(CP|T+)$? Why?

3. Have a look at the Bayes' Rule equation again. Write out the equation for the positive predictive value. Assuming $P(T+)$ doesn't change with prevalence,[1] what happens to the positive predictive value as the prevalence in the population increases? Do you find that reasonable or not? Explain.

4. Do you think the Ottawa ankle rules are useful in the medical field? Why or why not? How do your analyses affect your position?

---

[1] $P(T+)$ does, of course, change with prevalence, and this assumption is not necessary but simplifies the question. Your instructor may want to explain in more detail why it is not a required assumption.

Treatment Decision: Effective or Not? 

Name: ______________________________________   Section Number: __________

To be graded, all assignments must be completed and submitted on the original book page.

EXHIBIT 1

Pumpkin Powered Prostates _______________

Pumpkins and Prostate Health

Title: Pumpkin Seed Oil May Be a Halloween Treat

Author: Elena Conis

Source: *Los Angeles Times*, October 25, 2010, http://articles.
latimes.com/2010/oct/25/health/la-he-nutrition-lab-
pumpkin-20101025

In the News ...

According to the article in the *Los Angeles Times* by Elena Conis, for centuries pumpkin seeds have been a home remedy used to control or increase the frequency of urination in adults, children, and livestock. Knowing this about pumpkin seeds has prompted researchers to explore the possibility of a link between eating the seeds and better prostate health.

German researchers have been very involved in exploring this possible connection. The article summarized one study, the results of which were published in a German journal in 2000. According to the *Los Angeles Times*, the study:

> randomly selected among about 500 men to take either 1,000 milligrams of pumpkin seed oil extract or a placebo every day for 12 months. Symptoms improved in 65% of the men who took the oil, which the researchers interpreted as a promising (and statistically significant) result, even though symptoms also improved in 54% of the men who took the placebo.

Questions

1. What is being tested and how did it turn out?

2. Can we be certain that pumpkin seed oil is effective? Why or why not?

Stutter Stopper?

Drugs for Stutterers

Title: Drug for Stutterers Shows Promise: Indevus Says Pill Reduced Incidents for Most in 1st Trial

Author: Stephen Heuser

Source: *Boston Globe,* May 25, 2006, http://www.boston.com/yourlife/health/diseases/articles/2006/05/25/drug_for_stutterers_shows_promise/

The following is an extract from a *Boston Globe* article on stuttering:

A potential pill to treat stuttering took a step forward yesterday when Indevus Pharmaceuticals Inc. of Lexington said its experimental drug reduced stuttering in a majority of patients in its first clinical trial.

The 132-patient trial is the largest human test ever conducted on a drug for stuttering, according to the company.

…

The Indevus drug, called pagoclone, was given to 88 patients in escalating doses, with the rest of the trial subjects receiving a placebo. The patients were then tracked using several widely accepted measures of stuttering.

…

On a third rating scale, based on doctors' impressions, the pagoclone patients scored a "numerically superior rating" to the placebo group, but the finding did not reach statistical significance.

Questions

1. What is being tested, and how did it turn out?

2. Can we be certain that pagoclone is ineffective? Why or why not?

Statistical Significance in the Media—Part I

Name: ____   Section Number: _________

To be graded, all assignments must be completed and submitted on the original book page.

EXHIBIT 1

Rocky Biloba ________________________________

Drug Effectiveness

Title: Ginkgo Biloba Ineffective against Dementia, Researchers Find

Author: Roni Caryn Rabin

Source: *New York Times,* November 18, 2008, http://www.nytimes.com/2008/11/19/health/research/18gingko.html

The following is an extract from a *New York Times* article on ginkgo biloba:

> The largest and longest independent clinical trial to assess ginkgo biloba's ability to prevent memory loss has found that the supplement does not prevent or delay dementia or Alzheimer's disease, researchers are reporting.

> The study is the first trial large enough to accurately assess the plant extract's effect on the incidence of dementia, experts said, and the results dashed hopes that it is an effective preventative. In fact, there were more cases of dementia among participants who were taking ginkgo biloba than among those who were taking a placebo, though the difference was not statistically significant.

> "We were disappointed," said Dr. Steven T. DeKosky, dean of the School of Medicine at the University of Virginia and the principal investigator. "We were hopeful this would work."

Questions

1. Clearly identify (in your own well-chosen words) the two outcomes that are being compared. That is, what choice has to be made?

2. Which outcome was chosen? How is this related to the Type I Error Rate (false positive rate)?

Unofficially Dangerous?

Drug Side Effects

Title: Supreme Court Rules against Zicam Maker

Author: Adam Liptak

Source: *New York Times,* March 23, 2011, http://www.nytimes.com/2011/03/23/
health/23bizcourt.html

The following is an extract from a *New York Times* article on Zicam:

> The Supreme Court unanimously ruled on Tuesday that investors suing a drug company for securities fraud may rely on its failure to disclose scattered reports of adverse affects [sic] from an over-the-counter cold remedy that fell short of statistical significance.
>
> The case involved Zicam, a nasal spray and gel made by Matrixx Initiatives and sold as a homeopathic medicine. From 1999 to 2004, the plaintiffs said, the company received reports that the products might have caused some users to lose their sense of smell, a condition called anosmia.
>
> Matrixx did not disclose the reports and in 2003, the company said it was "poised for growth" and had "very strong momentum" though, by the plaintiffs' calculations, Zicam accounted for about 70% of its sales.
>
> ...
>
> In the case before the justices, Matrixx Initiatives Inc. v. Siracusano, No. 09-1156, lawyers for Matrixx argued that it should not have been required to disclose small numbers of unreliable reports, which were the only ones available in 2004, they said. They added that the company should face liability for securities fraud only if the reports had been collectively statistically significant.

Questions

1. This is an unusual situation. Clearly identify (in your own well-chosen words) the two outcomes that are being compared when the words "statistically significant" are used.

2. What is Matrixx claiming? How is their claim related to a Type I Error Rate (false positive rate)?

Statistical Significance in the Media—Part IV 

Name: ____   Section Number: _________

To be graded, all assignments must be completed and submitted on the original book page.

EXHIBIT 1

Dangerous Training? 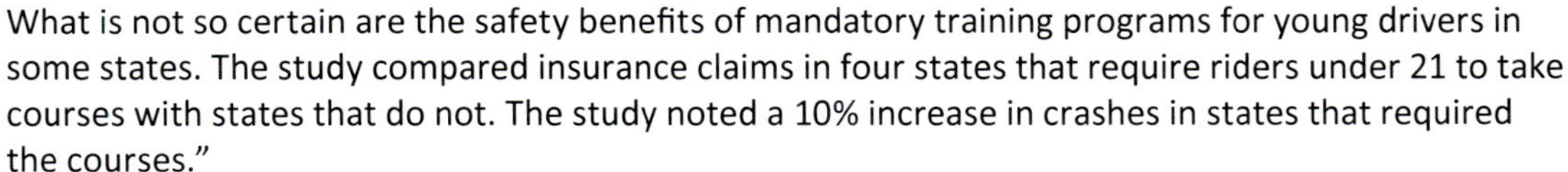

Crash Test

Title: Motorcycle Training Does Not Reduce Crash Risk, Study Says

Author: Cheryl Jensen

Source: *New York Times,* April 5, 2010, http://wheels.blogs.nytimes.com/2010/04/05/motorcycle-training-does-not-reduce-crash-risk-study-says/

Consider the following statements from a recent *New York Times* article discussing a recent study by the Highway Loss Data Institute regarding the effectiveness of training courses in improving motorcycle safety, and consider the questions below. The article says:

> What is not so certain are the safety benefits of mandatory training programs for young drivers in some states. The study compared insurance claims in four states that require riders under 21 to take courses with states that do not. The study noted a 10% increase in crashes in states that required the courses."

> But that finding wasn't "statistically significant," Ms. McCartt [senior vice president for research at the Insurance Institute] said. That means the increase might or might not be real, although the institute found it worth noting. "It is important that it was going in the opposite direction of what people would expect," she said.

Questions

1. Clearly identify what the null (H_0) and the alternative (H_A) hypotheses appear to be in the context of this article.

2. Articulate what risk was involved, based on the decision made in the article, when the null (H_0) was rejected (or not). You should use the concept of Type I error in your answer.

Vouching for Vouchers

Voucher Controversy

Title: White House Ignores Evidence of How D.C. School Vouchers Work

Author: Editorial Board Opinion

Source: *Washington Post,* March 29, 2011, http://www.washingtonpost.com/opinions/white-house-ignores-evidence-of-how-dc-school-vouchers-work/2011/03/29/AFFsnHyB_story.html

Consider the following statements from a recent *Washington Post* editorial discussing the Obama strongly worded dismissal of school vouchers. The article says:

> That dismissal might come as a surprise to Patrick J. Wolf, the principal investigator who helped conduct the rigorous studies of the D.C. Opportunity Scholarship Program and who has more than a decade of experience evaluating school choice programs.

> Here's what Mr. Wolf had to say about the program in Feb. 16 testimony to the Senate Committee on Homeland Security and Governmental Operations. "In my opinion, by demonstrating statistically significant experimental impacts on boosting high school graduation rates and generating a wealth of evidence suggesting that students also benefited in reading achievement, the DC OSP has accomplished what few educational interventions can claim: It markedly improved important education outcomes for low-income inner-city students."

Questions

1. The phrase "statistically significant" is used in Mr. Wolf's testimony. Clearly identify what the null (H_0) and alternative (H_A) hypotheses appear to be in the context of that testimony.

2. Articulate what risk was involved, based on the decision made in the article, when the null (H_0) was rejected (or not). You should use the concept of Type I Error Rate (false positive rate) in your answer.

Practical Significance versus Statistical Significance

Name: _______________________________________   Section Number: __________

To be graded, all assignments must be completed and submitted on the original book page.

EXHIBIT 1

Effect Size Matters

From *Time* Health

Women in the flibanserin group self-reported an average of 2.8 sexually satisfying events in the four-week baseline period. In the final four weeks of the 24-week study period, those women reported an average of 4.5 sexually satisfying events, a more than 50% increase. Women in the placebo group reported an average increase from 2.7 events to 3.7. The difference in effect between flibanserin and the placebo—about 0.8 sexually satisfying events—was statistically significant, the drug company said. The side effects from the drug, which included dizziness and fatigue, among others, were mild to moderate and transient.

Read more: http://www.time.com/time/health/article/0,8599,1939884,00.html

Questions

1. Explain where the "about 0.8 sexually satisfying events" statement comes from.

2. Read the description of the study carefully. What was the difference in effect between flibanserin and the placebo on an average weekly basis? Was the difference impressively large? Explain.

Catching a Breath

Stopping Sleep Apnea

Title: Statistical versus Clinical Significance: They Are Not the Same

Author: The Skeptical Scalpel

Source: *Skeptical Scalpel,* August 8, 2011, http://skepticalscalpel.blogspot.com/2011/08/ statistical-vs-clinical-significance.html

In reference to an article that appeared on MedPage Today, August 5th, 2011 ("Compression Stocking Help Sleep Apnea," by Michael Smith), the Skeptical Scalpel writes:

> MedPage Today featured an article about the beneficial effects of daytime wearing of compression stockings on obstructive sleep apnea. The premise was that increased edema in the neck could be caused by fluid coming from the legs when patients were in the supine position at night. Twelve patients who served as their own controls wore compression stockings for a week and then no stockings for a week alternating. The stockings lowered the amount of fluid in the neck by 60%, a statistically significant difference. So far, so good.

> This resulted in another highly statistically significant finding, which was a 36% reduction in episodes of apnea [cessation of breathing] and hypopnea [inadequate breathing]. Sounds good, right? The problem is that the average number of episodes of apnea/hypopnea decreased from 48 per hour to 31 per hour. Patients experiencing more than 30 episodes of apnea/hypopnea per hour are classified as having severe obstructive sleep apnea. This means that the treatment only put the patients in the low range of severe obstructive sleep apnea. They still would require maximum therapy. Is a reduction in apnea/hypopnea episodes that does not move the patient out of the severe category really clinically significant? It does not seem so to me.

Questions

1. Clearly identify what the null (H_0) and alternative (H_A) hypotheses appear to be in the context of this article. Which one was chosen and why?

2. The Skeptical Scalpel is making a point about practical significance. What is that point? Do you agree?

A Practical Discussion  

Name: ____   **Section Number:** _________

To be graded, all assignments must be completed and submitted on the original book page.

EXHIBIT 1

The Economics of No Significance _______

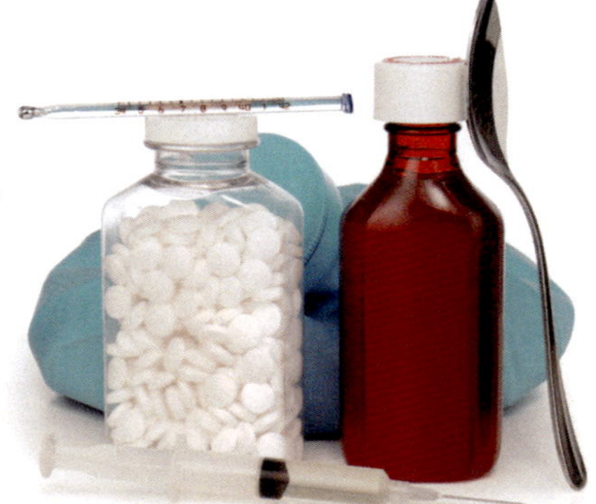

Title:	US Supreme Court: Statistical Significance Not Needed in Drug Lawsuits
Author:	Heidi Ledford
Source:	Nature.com, March 23, 2011. http://blogs.nature.com/ news/2011/03/us_supreme_court_statistical_s.html

Recall the example on pharmaceutical company Matrixx from Beyond the Numbers 3.3:

In a unanimous decision, the US Supreme Court ruled yesterday that a pharmaceutical company may be required to notify investors of safety reports regarding its products, even if those reports do not rise to the level of statistical significance.

Investors sued the company, based in Scottsdale, Arizona, arguing that it should have notified them earlier about reports that some of its popular zinc gluconate cold medications may have robbed some users of their sense of smell.

When news of the possible link finally became public, Matrixx stock plummeted. Matrixx tried to shoot down the lawsuit by arguing that the adverse event reports it received about Zicam were not statistically significant. But Judge Sonia Sotomayor, writing for the court, said that test would be too stringent. "Both medical experts and the Food and Drug Administration (FDA) rely on evidence other than statistically significant data to establish an inference of causation," she wrote. "It thus stands to reason that reasonable investors would act on such evidence."

Question

1. This is clearly a case wherein results that were not statistically significant were judged by the Supreme Court to be practically significant. What practical significance was in question here? Be sure to consider who brought the case against Matrixx.

The N Crowd

An MD/Ph.D at a major research institution is studying a new cannabis-based pain medication. The patients who volunteer for the study are randomly divided into two groups. Group 1 is given a placebo (Treatment 1). After one hour, group members rate the effectiveness of the pain relief on a scale of 1 to 100. Group 2 is given the new drug (Treatment 2), then group members rate the pain relief effectiveness in a similar way. Initially, only 25 volunteers were available for each group. By the end of the month, however, there were 25,000 volunteers in each group, making this the largest clinical trial in recent memory. Amazingly, there is only a one-point difference between the ratings of the placebo group and the ratings of the active treatment group at each stage. Formally, the following choice has to be made:

- H_0: Treatment 2 is no different than Treatment 1

- H_A: Treatment 2 is different than Treatment 1

Questions

1. Help the researcher decide if her results are statistically significant each week by filling out the last two rows of Table 3.18. The researcher adopted an alpha level of 0.05. Unless your instructor tells you otherwise, you may use the convenient GRAPHPAD® applet to complete your calculations. This applet is available at: http://www.graphpad.com/quickcalcs/ttest1/?Format=SD. The first table entry has already been filled out, but you may want to use that answer to make sure that you understand the process.

TABLE 3.18 Statistically Significant Results

	1st Week	2nd Week	3rd Week	4th Week
Sample size in each group (N)	25	250	2,500	25,000
Standard deviation in each group	20	20	20	20
Treatment 1 mean (placebo)	500	500	500	500
Treatment 2 mean (new drug)	501	501	501	501
p-value	0.8604			
Statistically significant? Check 'Yes' or 'No'	☐ Yes ☑ No	☐ Yes ☐ No	☐ Yes ☐ No	☐ Yes ☐ No

2. Formally, a p-value was the probability of seeing data which support H_A at least as strongly as what you have just produced, given H_0 is assumed to be true. In this exhibit, what kinds of data would support H_A? Give an example.

3. *Effect size* can be defined as the observed difference between two means, divided by the common standard deviation. What is the effect size for each of the four weeks?

4. What does this exercise have to say about the relationship between sample size and statistical significance? Be very specific in your answer.

5. What does this exercise have to say about whether one can reliably infer practical significance from statistical significance?

6. Do you think a study's sample size can ever be too large? Explain your reasoning.

Accept or Fail to Reject? 
Semantics or Real?

Name: ____ Section Number: __________

To be graded, all assignments must be completed and submitted on the original book page.

EXHIBIT 1

Legally Speaking

Title: Innocent versus Not Guilty: Jury Decision Based Entirely on Evidence

Author: Hugh Duvall

Source: http://www.defendingoregon.com/innocent-v-guilty/

Many attorneys have written about the difference between innocence and non-guilt. Most quickly turn to complex legalese that would obscure our point here. This article paints a particularly clear picture of the use, however:

> Juries never find defendants innocent. They cannot. Not only is it not their job, it is not within their power. They can only find them "not guilty." Once a person has been charged with having committed a crime, there is no mechanism by which that individual can prove his innocence. Yes, the law provides that the person is innocent unless proven guilty, but that is a legalism. It is not, nor could it be, a factual statement. The person, in fact, did or did not commit an offense. Each time a member of the media or other citizen states that William Kennedy Smith or one of the officers accused of beating Rodney King was found "innocent," they are not only incorrect, but are also ingraining within potential jurors a misconception about their role. They enhance the risk that enough jurors on a panel will retire into a jury room believing that it is their task to determine whether there is enough evidence to find a defendant innocent.

Questions

1. The role of the prosecution is to establish guilt beyond a reasonable doubt. The defense has no such burden of proof. How does this affect the ability of a jury to find a defendant innocent?

2. If "I" denotes the event "defendant is presumed innocent" and "E" denotes "evidence presented by prosecution," then a juror's job is to evaluate P(E|I).* This probability can be ascertained to be big or small, depending on the case. Identify which of these (big or small) leads logically to a conclusion of guilty and explain why the other can't logically lead to a conclusion of innocence.

3. In this article, the author states that, "As a society, in administering the prosecution function, we must keep at the forefront of our mind that there is no way to reverse the implication of charging someone with a crime. Allowing ourselves to ignore the distinction between a jury's ability to find someone 'not guilty' and its inability to find someone 'innocent' works against this important interest." Explain in your own words what this means and describe why it is important in this discussion of "innocent" versus "not guilty."

4. Does this example change your perception of innocence and guilt? Does it change your perception of the legal system? Explain your answer.

* See Beyond the Numbers 3.1 for a definition of this notation if you are unfamiliar with it.

Alternative Evidence

Title: "It's Like … You Know": The Use of Analogies and Heuristics in Teaching Introductory Statistical Methods

Author: Michael A. Martin, Australian National University

Source: *Journal of Statistics Education* Volume 11, Number 2 (2003). www.amstat.org/publications/jse/v11n2/martin.html.

Analogies between the United States criminal justice system and hypothesis testing are very useful to the study of statistics. How so? They make it easier to understand the logic of an abstract task that is often seen as only having deductive relevance. In this article, Professor Michael Martin created a number of criminal trial analogies to explain statistics' most common (and most commonly confusing) concepts:

Criminal Trial Concepts	Hypothesis Testing Concepts
1. Defendant is innocent	Null hypothesis
2. Defendant is guilty	Alternative hypothesis
3. Verdict is to acquit	Failure to reject the null hypothesis
4. Verdict is to convict	Rejection of the null hypothesis
5. Presumption of innocence	Assumption that the null hypothesis is true
6. Conviction of an innocent person	Type I error
7. Acquittal of a guilty person	Type II error
8. Beyond reasonable doubt	Fixed (small) probability of Type I error

Questions

1. Can you explain each of these eight correspondences? Take each analogy in turn, and explain how its components are analogous. You must use the language of this course in your answers.

2. Is the distinction between accepting and failing to reject a null hypothesis a real distinction, or just semantics? Defend your answer based on what you have learned from these Exhibits.

3. Use a hobby, interest, job, or event from your own life to create your own analogy to one or more of the key hypothesis testing concepts listed in Exhibit 2.

Origins of Power  

Name: ____ Section Number: __________

To be graded, all assignments must be completed and submitted on the original book page.

Background

We haven't talked much about Type II errors since we started hypothesis testing. In hypothesis testing, the Type II error rate is the empirical probability of failing to reject H_0 when you should. This rate typically denoted by the Greek letter β. The power of the hypothesis test, $1-\beta$, is directly analogous to sensitivity. Think of it as the probability of choosing H_A when H_A is the right choice.

The computation of power can be a complex endeavor, even for elementary forms of H_0 and H_A. However, we can gain valuable insights into the power of a statistical test by using a freely-available online tool to handle all of the complex computations for us.

EXHIBIT

Power and Beauty

Title:	Power and Sample Size.com
Author:	HyLown Consulting, LLC
Source:	http://powerandsamplesize.com/Calculators/Test-1-Mean/1-Sample-Equality

Suppose you are interested in how people rate their looks on a 20 point scale, with a score of 0 meaning *unbearably ugly* and a score of 20 meaning *hopelessly gorgeous.* Let's suppose you want to test the following hypothesis:

- H_0: true population average is 10

- H_A: true population average is not 10

What we want to do is to answer the question:

"How likely is it that we will fail to reject H_0 if H_0 is truly false?"

The answer to this question leads us to the Type II error rate, from which power is easily computed. To come up with an answer, we first have to ask:

"What is the true average if H_0 is false?"

H_A only tells us, in this case, that the true average is something other than 10. We will begin by considering four possible averages that are different from 10: 10.5, 11, 11.5, and 12.

Questions

1. Access the applet at the web address listed in the Source line above. When the applet page loads:

    - Select "Power"

    - Enter the sample size (e.g. 50)

    - Enter True Mean (e.g. 11)

    - Enter Hypothesized Mean (always 10 for this problem)

    - Enter the Standard Deviation as 5

    - Leave the alpha level (Type I error rate) fixed at 5%

    - Hit Calculate

    - Record the Power value

    - Do this for all True Mean and Sample Size combinations shown in the table below. One row is already done for you. Make sure you can confirm those entries and then finish finding the rest of the entries in the table. It is safest to reload the page between each calculation.

TABLE 3.19 Power Value Results

Power Table	Possible Real Values of the Population Average			
Sample Size	10.5	11	11.5	12
10	A	A	A	A
50	0.11	0.29	0.57	0.81
100	B	B	B	B
1,000	C	C	C	C
10,000	D	D	D	D

a. The correct values in the first column (under the True Mean of 10.5) are:

   i.   A = 0.10; B = 0.52; C = 1.00; D = 1.00

   ii. A = 0.06; B = 0.17; C = 0.89; D = 1.00

   iii. A = 0.25; B = 0.98; C = 1.00; D = 1.00

   iv. A = 0.16; B = 0.85; C = 1.00; D = 1.00

b. The correct values in the second column (under the True Mean of 11) are:

   i.   A = 0.10; B = 0.52; C = 1.00; D = 1.00

   ii. A = 0.06; B = 0.16; C = 0.89; D = 1.00

   iii. A = 0.25; B = 0.98; C = 1.00; D = 1.00

   iv. A = 0.16; B = 0.85; C = 1.00; D = 1.00

c. The correct values in the third column (under the True Mean of 11.5) are:

    i.   A = 0.10; B = 0.52; C = 1.00; D = 1.00

    ii. A = 0.06; B = 0.16; C = 0.89; D = 1.00

    iii. A = 0.25; B = 0.98; C = 1.00; D = 1.00

    iv. A = 0.16; B = 0.85; C = 1.00; D = 1.00

d. The correct values in the fourth column (under the True Mean of 12) are:

    i.   A = 0.10; B = 0.52; C = 1.00; D = 1.00

    ii. A = 0.06; B = 0.16; C = 0.89; D = 1.00

    iii. A = 0.25; B = 0.98; C = 1.00; D = 1.00

    iv. A = 0.16; B = 0.85; C = 1.00; D = 1.00

e. Look at the table. How does power change as sample size changes, regardless of the true average?

f. Look at the table. How does power change as the possible true average changes, regardless of sample size?

g. What would happen to power if you changed alpha? Investigate that question using the online calculator that you used to fill out the table above. Explain what you find. Think back to how sensitivity and specificity behaved in screening tests. How is this similar?

Appendix

Expressions and Formulas

Throughout the use of this book, you will find these expressions and formulas to be useful.

Module 1

SAMPLE PROPORTION

This simple formula for a sample proportion, where n is the sample size and k is the number of "positive" outcomes (e.g., "Yes" answers).

$$\hat{p} = \frac{k}{n}$$

SAMPLE MEAN

$$\bar{x} = \frac{x_1 + x_2 + \ldots + x_n}{n}$$

SAMPLE MEDIAN

Arrange all data points in increasing order. If there are an odd number of data points, the median is the one in the middle. If there are an even number of data points, the median is the mean of the two in the middle.

SAMPLE STANDARD DEVIATION

$$s = \sqrt{\frac{(x_1 - \bar{x})^2 + (x_2 - \bar{x})^2 + \ldots + (x_n - \bar{x})^2}{n - 1}}$$

SAMPLE VARIANCE

The sample variance is just the square of the same standard deviation.

$$r = \frac{n(\sum xy) - (\sum x)(\sum y)}{\sqrt{[n\sum x^2 - (\sum x)^2][n\sum y^2 - (\sum y)^2]}}$$

CORRELATION COEFFICIENT

Given n pairs of observations in (x, y):

Module 2

95% MARGIN OF ERROR FOR POPULATION PROPORTION BASED ON SRS OF SIZE n

$$\text{MOE} = \sqrt{\frac{1}{n}}$$

MARGIN OF ERROR FOR OTHER CONFIDENCE LEVELS, FOR POPULATION PROPORTION BASED ON SRS OF SIZE *n*

$$\text{MOE} = \frac{z^*}{2}\sqrt{\frac{1}{n}}\text{ , where }z^*\text{ is chosen from the table below, as appropriate.}$$

Level of Confidence	z^*	Level of Confidence	z^*
50%	0.67	90%	1.64
60%	0.84	95%	2.00
70%	1.04	99%	2.58
80%	1.28	99.9%	3.29

CONFIDENCE INTERVAL FOR POPULATION PROPORTION BASED ON SRS OF SIZE *n*

$$\hat{p} \pm \frac{z^*}{2}\sqrt{\frac{1}{n}}\text{ , where }z^*\text{ is chosen from the table below, as appropriate.}$$

Level of Confidence	z^*	Level of Confidence	z^*
50%	0.67	90%	1.64
60%	0.84	95%	2.00
70%	1.04	99%	2.58
80%	1.28	99.9%	3.29

THE EMPIRICAL RULE

Empirical Rule:

Suppose a bell-shaped distribution has a mean μ and a standard deviation σ. Then:

a. About 68% of all observations represented by that distribution will fall within one standard deviation of the mean.

b. About 95% will fall within two standard deviations of the mean.

c. About 99.7% will fall within three standard deviations of the mean.

Graphically:

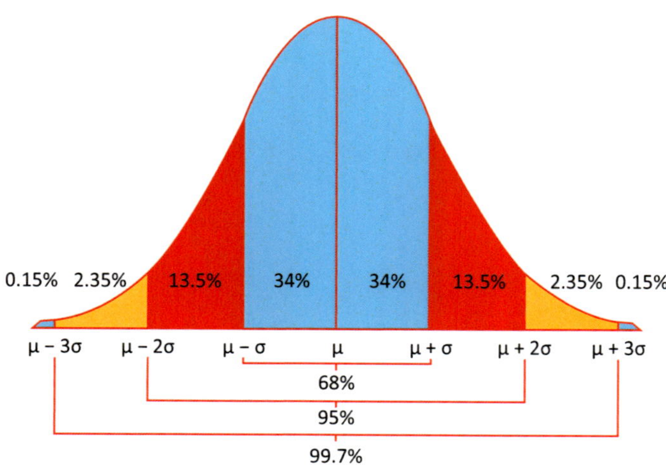

THE EMPIRICAL RULE APPLIED TO SAMPLING DISTRIBUTION OF SAMPLE PROPORTION

If you compute the sample proportion based on an SRS of size *n*, then the following graphic describes how all such sample proportions would be distributed.

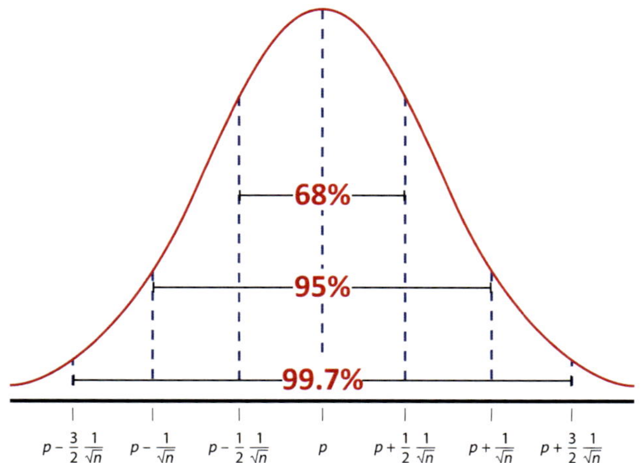

CONFIDENCE INTERVAL FOR POPULATION MEAN BASED ON AN SRS OF SIZE *n*

$\bar{x} \pm z^* \dfrac{s}{\sqrt{n}}$, where *s* is the sample standard deviation, and z^* is chosen from the table below, as appropriate.

Level of Confidence	z^*	Level of Confidence	z^*
50%	0.67	90%	1.64
60%	0.84	95%	2.00
70%	1.04	99%	2.58
80%	1.28	99.9%	3.29

Module 3

BAYES' RULE

The notation P(A|B) is read as "the probability of A, given B, has occurred." So the "|" symbol is read as "given." Formally, A and B are called *events* and P(A|B) is a *conditional* probability.

Bayes' rule is a very useful way of relating conditional and unconditional probabilities. According to this rule, for any two events A and B, we have:

$$P(A|B) = \frac{P(B|A) \times P(A)}{P(B)}$$

OVERALL ACCURACY

Overall Accuracy is defined as (Prevalence) × Sensitivity + (1 − Prevalence) × Specificity

STANDARD SCORE FOR TESTING A ONE-SAMPLE HYPOTHESIS ABOUT A POPULATION PROPORTION

Give a sample proportion $\hat{p}$ and a hypothesized value of p, p_0, use the following:

$$z = \frac{(\hat{p} - p_0)}{\sqrt{\dfrac{(p_0)(1 - p_0)}{n}}}$$

The way in which this is used depends on the form of the hypothesis.

USING A CONFIDENCE INTERVAL TO TEST A TWO-TAILED HYPOTHESIS FOR A POPULATION PROPORTION

To test the following hypothesis with a confidence interval, follow the steps below:

$$H_0: p = p_0$$

$$H_A: p \neq p_0$$

Step 1: Establish a value for alpha level, the Type I error rate (typically $\alpha = 0.05$) and divide that by 2.

Step 2: Find the standard score from a standard score table (like the one included in this workbook) that corresponds to $\alpha/2$, Call this z^*. If $\alpha = 0.05$, then $z^* = 1.96$.

Step 3: Form the confidence interval:

$$\hat{p} \pm z^* \sqrt{\frac{(p_0)(1 - p_0)}{n}}$$

Step 4: See if p_0 is within the interval. If it is not, you can reject H_0. p-values do not come in play when using confidence intervals for testing.

TESTING A TWO-TAILED HYPOTHESIS FOR A POPULATION MEAN

To test the following hypothesis, follow the steps below:

$$H_0: \mu = \mu_0$$

$$H_A: \mu \neq \mu_0$$

Step 1: Establish a value for alpha level, the Type I error rate (typically $\alpha = 0.05$).

Step 2: Compute the standard score (z-value):

$$z = \frac{(\bar{x} - \mu_0)}{\dfrac{s}{\sqrt{n}}}$$

Step 3: Take the absolute value of z to the standard score table in this workbook (the same way you would for the > alternative) and record that value. The p-value for the two-tailed alternative is twice this value.

Step 4: Compare the two-tailed p-value to α, the preset Type I error rate. If the p-value is smaller, reject H_0; if it is larger, fail to reject H_0.

Tables

VALIDATION OF THE STANDARDIZED FIELD SOBRIETY TEST BATTERY AT BACS BELOW 0.10%: DATA SET

Case	HGN	OLS	WAT	Total FST	Actual BAC	Case	HGN	OLS	WAT	Total FST	Actual BAC
229	0	0	1	1	0	130	2	1	1	4	0.07
254	0	0	1	1	0.02	297	4			4	0.08
66		1		1	0.067	271	2	0	2	4	0.1
142	2	0	0	2	0.005	119		2	2	4	0.121
217	2	0	0	2	0.03	296	2	1	2	5	0.01
191	2			2	0.034	88	2	1	2	5	0.02
182	2	0	0	2	0.038	176	4	1	0	5	0.02
109	2			2	0.04	295	2	1	2	5	0.02
259	2	0	0	2	0.04	137		2	3	5	0.03
199	0	1	1	2	0.048	288	2	1	2	5	0.03
113	2	0	0	2	0.05	13	0	1	4	5	0.043
22	2			2	0.06	54	4	1	0	5	0.046
67	2		1	3	0.022	20	2		3	5	0.049
145	0	3	0	3	0.03	164	2	1	2	5	0.07
53	2	0	1	3	0.032	99	2	2	1	5	0.12
15	2	0	1	3	0.04	223			5	5	0.18
287	2	0	1	3	0.04	91	2	2	2	6	0.017
89	2	1	0	3	0.05	111	2	3	1	6	0.027
123	0	1	2	3	0.05	31	2	3	1	6	0.039
258	2	1	0	3	0.053	29	4	1	1	6	0.05
35	2	0	2	4	0	76	4	1	1	6	0.05
11	2	1	1	4	0.01	260	6			6	0.07
247	4			4	0.016	86	6			6	0.088
6	2	0	2	4	0.02	8	6			6	0.09
294	2	0	2	4	0.02	212	6			6	0.11
231	2	0	2	4	0.03	114	4	1	1	6	0.116
74	2	1	1	4	0.04	79	6			6	0.12
214	2	1	1	4	0.04	198	6			6	0.2
58	4			4	0.05	59	6			6	0.22
14	2	1	1	4	0.058	107	6			6	0.23
34	4	0	0	4	0.058	221	6			6	0.23
12	2	0	2	4	0.06	24	6			6	0.26
211	2	1	1	4	0.06	157	6			6	0.33
232	4	0	0	4	0.06	126	4	1	2	7	0.04
293	2	2	0	4	0.06	289	2	1	4	7	0.04

Case	HGN	OLS	WAT	Total FST	Actual BAC
75	2	4	1	7	0.048
52	4	2	1	7	0.05
2	2	2	3	7	0.06
263	2	3	2	7	0.07
149	6	1		7	0.13
117		3	4	7	0.14
165	2	3	3	8	0.02
27	4	2	2	8	0.05
219	2	3	3	8	0.07
213	4	2	2	8	0.08
291	2	4	2	8	0.08
131	4	2	2	8	0.082
72	4	2	2	8	0.09
166	6	1	1	8	0.09
292	4	2	2	8	0.09
125	4	0	4	8	0.091
280	6	1	1	8	0.1
134	6	1	1	8	0.102
62	6	1	1	8	0.15
56	6	1	1	8	0.16
189	6	1	1	8	0.173
71	2	3	3	8	0.19
84	2	5	2	9	0
104	3	3	3	9	0.037
183	6	1	2	9	0.04
187	2	6	1	9	0.05
226	6	1	2	9	0.058
121	6	2	1	9	0.06
186	4	4	1	9	0.063
106	6	2	1	9	0.08
140	4	4	1	9	0.08
163	6	3	0	9	0.08
193	2	3	4	9	0.1
10	6	3		9	0.12
77	6	3		9	0.156
93	6		3	9	0.2
47	6		3	9	0.25
209	6	2	1	9	0.32
122	4	2	4	10	0.017

Case	HGN	OLS	WAT	Total FST	Actual BAC
282	6	4		10	0.08
112	6	1	3	10	0.09
37	6	1	3	10	0.1
116	6	1	3	10	0.1
234	4	2	4	10	0.11
124	6	3	1	10	0.142
115	6	4		10	0.21
227	4	4	3	11	0.06
185	6	3	2	11	0.07
224	4	4	3	11	0.07
135	3	3	5	11	0.078
28	6	2	3	11	0.08
50	4	4	3	11	0.08
298	6	3	2	11	0.085
152	6	2	3	11	0.104
205	6	3	2	11	0.11
96	6	3	2	11	0.12
144	6	3	2	11	0.12
200	6	1	4	11	0.14
257	6	3	2	11	0.14
286	6	4	1	11	0.15
159	6	1	4	11	0.16
133	6	3	2	11	0.176
80	6	3	2	11	0.3
30	4	5	3	12	0.05
16	6	1	5	12	0.069
32	6	4	2	12	0.076
1	4	2	6	12	0.089
235	6	3	3	12	0.09
204	6	3	3	12	0.11
220	6	3	3	12	0.11
238	6	4	2	12	0.11
23	6	4	2	12	0.12
97	6	3	3	12	0.12
101	6	3	3	12	0.12
281	6	4	2	12	0.12
188	6	5	1	12	0.128
70	6	4	2	12	0.131
38	6	3	3	12	0.14

Case	HGN	OLS	WAT	Total FST	Actual BAC	Case	HGN	OLS	WAT	Total FST	Actual BAC
127	6	2	2	10	0.028	128	6	4	4	14	0.1
178	6	0	4	10	0.07	208	4	7	3	14	0.1
57	6	4	2	12	0.14	262	6	3	5	14	0.1
85	6	4	2	12	0.14	4	6	4	4	14	0.12
206	6	4	2	12	0.14	103	6	3	5	14	0.12
240	6	3	3	12	0.17	242	6	4	4	14	0.12
250		4	8	12	0.18	5	6	5	3	14	0.13
170	6	2	4	12	0.19	64	6	4	4	14	0.13
94	6	6		12	0.22	261	6	4	4	14	0.14
132	6	4	2	12	0.27	268	6	5	3	14	0.14
148	6	7	0	13	0.05	78	6	4	4	14	0.15
129	4	7	2	13	0.07	181	6	4	4	14	0.15
252	6	4	3	13	0.08	190	6	3	5	14	0.15
120	6	4	3	13	0.09	194	6	5	3	14	0.15
241	6	4	3	13	0.09	82	6	1	7	14	0.16
278	6	4	3	13	0.1	255	6	3	5	14	0.16
65	6	3	4	13	0.11	83	6	4	4	14	0.2
154	4	6	3	13	0.11	108	6	5	3	14	0.2
201	4		9	13	0.11	158	6	5	3	14	0.22
87	6	7		13	0.12	207	6	3	5	14	0.28
167	6	4	3	13	0.12	55	6	6	3	15	0.1
19	6	4	3	13	0.13	173	6	6	3	15	0.1
98	6	5	2	13	0.132	230	6	4	5	15	0.1
17	6	4	3	13	0.14	249	6	6	3	15	0.1
46	6	3	4	13	0.14	279	6	4	5	15	0.11
228	6	5	2	13	0.14	45	6	3	6	15	0.12
233	6	4	3	13	0.14	184	6	5	4	15	0.12
237	6	4	3	13	0.16	51	6	3	6	15	0.13
102	6	3	4	13	0.17	44	6	5	4	15	0.14
68	4	6	3	13	0.19	143	6	4	5	15	0.15
277	6	4	3	13	0.19	265	6	4	5	15	0.15
284	6	3	4	13	0.19	100	6	5	4	15	0.17
81	6	3	4	13	0.2	285	6	5	4	15	0.17
162	6	3	4	13	0.23	275	6	4	5	15	0.2
175	4	6	4	14	0.07	273	6	4	5	15	0.22
196	6	2	6	14	0.074	171	6	4	5	15	0.23
153	6	3	5	14	0.098	40	6	3	6	15	0.24
105	6	4	4	14	0.1	33	6	9	1	16	0.11

Case	HGN	OLS	WAT	Total FST	Actual BAC	Case	HGN	OLS	WAT	Total FST	Actual BAC
156	6	7	3	16	0.11	172	6	7	5	18	0.14
266	6	7	3	16	0.11	49	6	4	8	18	0.15
18	6	6	4	16	0.12	216	6	7	5	18	0.15
169	6	7	3	16	0.12	270	6	8	4	18	0.15
174	6	7	3	16	0.13	195	6	7	5	18	0.16
251	6	4	6	16	0.14	256	6	6	6	18	0.16
264	6	4	6	16	0.14	239	6	4	8	18	0.17
7	6	7	3	16	0.15	276	6	4	8	18	0.19
25	6	6	4	16	0.16	210	6	4	8	18	0.2
218	6	4	6	16	0.19	110	6	4	8	18	0.21
244	6	5	5	16	0.21	95	6	4	8	18	0.23
236	6	5	5	16	0.22	168	6	4	8	18	0.24
141	6	7	4	17	0.1	272	6	4	8	18	0.27
9	6	6	5	17	0.11	36	6	5	8	19	0.14
26	6	5	6	17	0.12	179	6	6	7	19	0.14
222	6	6	5	17	0.12	243	6	9	4	19	0.15
147	6	5	6	17	0.13	63	6	5	8	19	0.16
225	6	6	5	17	0.13	139	6	8	5	19	0.16
269	6	7	4	17	0.17	203	6	7	6	19	0.17
290	6	4	7	17	0.17	160	6	7	6	19	0.19
42	6	5	6	17	0.18	215	6	7	6	19	0.19
146	6	6	5	17	0.18	202	6	6	7	19	0.22
69	6	5	6	17	0.2	3	6	4	9	19	0.23
177	6	3	8	17	0.2	118	6	9	4	19	0.23
136	6	6	5	17	0.219	90	6	12	2	20	0.12
274	6	4	7	17	0.24	245	6	6	8	20	0.12
92	6	4	7	17	0.26	43	4	8	8	20	0.13
161	6	3	8	17	0.29	155	6	9	5	20	0.165
61	6	6	5	17	0.32	138	6	7	7	20	0.19
246	6	9	3	18	0.07	253	6	7	8	21	0.17
39	6	7	5	18	0.1	41	6	8	7	21	0.22
21	6	7	5	18	0.11	283	6	7	8	21	0.24
48	6	8	4	18	0.12	192	6	9	7	22	0.14
180	6	7	5	18	0.13	151	6	8	8	22	0.18
267	6	5	7	18	0.13	248	6	9	8	23	0.23
73	6	9	3	18	0.14	197	6	12	8	26	0.17

CANCER SURVIVAL DATA TABLE

This table is referenced throughout Module 3 of this workbook. Please reference as necessary to complete the related Beyond the Numbers. Please note that there are more data sets located at www.statconcepts.com/datasets.

Stomach Cancer	Bronchus Cancer	Colon Cancer	Ovarian Cancer	Breast Cancer
124	81	248	1,234	1,235
42	461	377	89	24
25	20	189	201	1,581
45	450	1,843	356	1,166
412	246	180	2,970	40
51	166	537	456	727
1,112	63	519		791
46	64	455		1,804
103	155	406		3,460
876	859	365		719
146	151	942		
340	166	776		
396	37	372		
	223	163		
	138	101		
	72	20		
	245	283		

Example: If a standard score, *z*, is computed to be 1.73, then one would locate 1.73 in the table (see highlighting) and corresponding *p*-value would be 0.04182 ≈ 0.04. This would then have to be compared to the preset Type I error rate, often taken to be 0.05. This process only applies for simple hypotheses with a positive *z* score and a ">" in the alternative. Your instructor will show you how to use this table when *z* is negative and/or when the alternative is a "≠."

z	0	0.01	0.02	0.03	0.04	0.05	0.06	0.07	0.08	0.09
0	0.5	0.49601	0.49202	0.48803	0.48405	0.48006	0.47608	0.4721	0.46812	0.46414
0.1	0.46017	0.4562	0.45224	0.44828	0.44433	0.44038	0.43644	0.43251	0.42858	0.42465
0.2	0.42074	0.41683	0.41294	0.40905	0.40517	0.40129	0.39743	0.39358	0.38974	0.38591
0.3	0.38209	0.37828	0.37448	0.3707	0.36693	0.36317	0.35942	0.35569	0.35197	0.34827
0.4	0.34458	0.3409	0.33724	0.3336	0.32997	0.32636	0.32276	0.31918	0.31561	0.31207
0.5	0.30854	0.30503	0.30153	0.29806	0.2946	0.29116	0.28774	0.28434	0.28096	0.2776
0.6	0.27425	0.27093	0.26763	0.26435	0.26109	0.25785	0.25463	0.25143	0.24825	0.2451
0.7	0.24196	0.23885	0.23576	0.2327	0.22965	0.22663	0.22363	0.22065	0.2177	0.21476
0.8	0.21186	0.20897	0.20611	0.20327	0.20045	0.19766	0.19489	0.19215	0.18943	0.18673
0.9	0.18406	0.18141	0.17879	0.17619	0.17361	0.17106	0.16853	0.16602	0.16354	0.16109
1	0.15866	0.15625	0.15386	0.15151	0.14917	0.14686	0.14457	0.14231	0.14007	0.13786
1.1	0.13567	0.1335	0.13136	0.12924	0.12714	0.12507	0.12302	0.121	0.119	0.11702
1.2	0.11507	0.11314	0.11123	0.10935	0.10749	0.10565	0.10383	0.10204	0.10027	0.09853
1.3	0.0968	0.0951	0.09342	0.09176	0.09012	0.08851	0.08692	0.08534	0.08379	0.08226
1.4	0.08076	0.07927	0.0778	0.07636	0.07493	0.07353	0.07215	0.07078	0.06944	0.06811
1.5	0.06681	0.06552	0.06426	0.06301	0.06178	0.06057	0.05938	0.05821	0.05705	0.05592
1.6	0.0548	0.0537	0.05262	0.05155	0.0505	0.04947	0.04846	0.04746	0.04648	0.04551
1.7	0.04457	0.04363	0.04272	0.04182	0.04093	0.04006	0.0392	0.03836	0.03754	0.03673
1.8	0.03593	0.03515	0.03438	0.03362	0.03288	0.03216	0.03144	0.03074	0.03005	0.02938
1.9	0.02872	0.02807	0.02743	0.0268	0.02619	0.02559	0.025	0.02442	0.02385	0.0233
2	0.02275	0.02222	0.02169	0.02118	0.02068	0.02018	0.0197	0.01923	0.01876	0.01831
2.1	0.01786	0.01743	0.017	0.01659	0.01618	0.01578	0.01539	0.015	0.01463	0.01426
2.2	0.0139	0.01355	0.01321	0.01287	0.01255	0.01222	0.01191	0.0116	0.0113	0.01101
2.3	0.01072	0.01044	0.01017	0.0099	0.00964	0.00939	0.00914	0.00889	0.00866	0.00842
2.4	0.0082	0.00798	0.00776	0.00755	0.00734	0.00714	0.00695	0.00676	0.00657	0.00639
2.5	0.00621	0.00604	0.00587	0.0057	0.00554	0.00539	0.00523	0.00508	0.00494	0.0048
2.6	0.00466	0.00453	0.0044	0.00427	0.00415	0.00402	0.00391	0.00379	0.00368	0.00357
2.7	0.00347	0.00336	0.00326	0.00317	0.00307	0.00298	0.00289	0.0028	0.00272	0.00264
2.8	0.00256	0.00248	0.0024	0.00233	0.00226	0.00219	0.00212	0.00205	0.00199	0.00193
2.9	0.00187	0.00181	0.00175	0.00169	0.00164	0.00159	0.00154	0.00149	0.00144	0.00139
3	0.00135	0.00131	0.00126	0.00122	0.00118	0.00114	0.00111	0.00107	0.00104	0.001
3.1	0.00097	0.00094	0.0009	0.00087	0.00084	0.00082	0.00079	0.00076	0.00074	0.00071
3.2	0.00069	0.00066	0.00064	0.00062	0.0006	0.00058	0.00056	0.00054	0.00052	0.0005
3.3	0.00048	0.00047	0.00045	0.00043	0.00042	0.0004	0.00039	0.00038	0.00036	0.00035
3.4	0.00034	0.00032	0.00031	0.0003	0.00029	0.00028	0.00027	0.00026	0.00025	0.00024
3.5	0.00023	0.00022	0.00022	0.00021	0.0002	0.00019	0.00019	0.00018	0.00017	0.00017
3.6	0.00016	0.00015	0.00015	0.00014	0.00014	0.00013	0.00013	0.00012	0.00012	0.00011

Three Sample Rubrics for Assessing Group Work

PEER EVALUATION FORM FOR GROUP WORK

Name: ____   Section Number: _________

Write the name of each of your group members in a separate column. For each person, indicate the extent to which you agree with the statement on the left, using a scale of 1–4 (1 = strongly disagree; 2 = disagree; 3 = agree; 4 = strongly agree). Total the numbers in each row.

Evaluation Criteria	Yourself	Group Member:	Group Member:	Group Member:	Group Member:	Group Member:
Attends Group Meetings Regularly and Arrives on Time						
Contributes Meaningfully to Group Discussions						
Completes Group Assignments on Time						
Prepares Work in a Quality Manner						
Demonstrates a Cooperative and Supportive Attitude						
Contributes Significantly to the Success of the Project						
Totals						

1. How effectively did your group work?

2. Were the behaviors of any of your team members particularly valuable or detrimental to the team? Explain.

3. What did you learn about working in a group from this project that you will carry into your next group experience?

Adapted from a peer evaluation form development at Johns Hopkins University (October, 2006).

RUBRIC FOR ASSESSING GROUP MEMBERS' ABILITY TO PARTICIPATE EFFECTIVELY AS PART OF A TEAM

Rater: ____ Date:________________

Group Topic: ____

Circle the appropriate score for each criterion for each member of your group.

Member Rated (Be sure to rate yourself, too!)	Listening Skills	Openness to Others' Ideas	Preparation	Contribution	Leadership
	0 1 2 3 4 5	0 1 2 3 4 5	0 1 2 3 4 5	0 1 2 3 4 5	0 1 2 3 4 5
	0 1 2 3 4 5	0 1 2 3 4 5	0 1 2 3 4 5	0 1 2 3 4 5	0 1 2 3 4 5
	0 1 2 3 4 5	0 1 2 3 4 5	0 1 2 3 4 5	0 1 2 3 4 5	0 1 2 3 4 5
	0 1 2 3 4 5	0 1 2 3 4 5	0 1 2 3 4 5	0 1 2 3 4 5	0 1 2 3 4 5
	0 1 2 3 4 5	0 1 2 3 4 5	0 1 2 3 4 5	0 1 2 3 4 5	0 1 2 3 4 5
	0 1 2 3 4 5	0 1 2 3 4 5	0 1 2 3 4 5	0 1 2 3 4 5	0 1 2 3 4 5

Criterion	Excellent (5)	Good (4)	Fair (3)	Needs to Improve (2)	Unacceptable (1)	Missing (0)
Listening Skills	Routinely restates what others say before responding; rarely interrupts; frequently solicits others' contributions; sustains eye contact	Often restates what others say before responding; usually does not interrupt; often solicits others' contributions; makes eye contact	Sometimes restates what others say before responding; sometimes interrupts; sometimes asks for others' contributions; sometimes makes eye contact	Rarely restates what others say before responding; often interrupts; rarely solicits others' contributions; does not make eye contact; sometimes converses with others when another team is speaking	Doesn't restate what others say before responding; often interrupts; doesn't ask for contributions from others; is readily distracted; often talks with others when another team member speaks	Never shows up and never contributes
Openness to Others' Ideas	Listens to others' ideas without interrupting; responds positively to ideas even if rejecting; asks questions about the ideas	Listens to others' ideas without interrupting; responds positively to the ideas even when rejecting	Sometimes listens to others' ideas without interrupting; generally responds to the ideas	Interrupts others' articulation of their ideas; does not comment on the ideas	Interrupts others' articulation of their ideas; makes deprecatory comments and/or gestures	Never shows up and never contributes
Preparation	Always completes assignments; always comes to team sessions with necessary documents and materials; does additional research, reading, writing, designing, implementing	Typically completes assignments; typically comes to team sessions with necessary documents and materials	Sometimes completes assignments; sometimes comes to team sessions with necessary documents and materials	Sometimes completes assignments; sometimes comes to team sessions with necessary documents and materials	Typically does not complete assignments; typically comes to team sessions without necessary documents and materials	Never shows up and never contributes
Contribution	Always contributes; quality of contributions is exceptional	Usually contributes; quality of contributions is solid	Sometimes contributes; quality of contributions is fair	Sometimes contributes; quality of contribution is inconsistent	Rarely contributes; contributions are often peripheral or irrelevant; frequently misses team sessions	Never shows up and never contributes
Leadership	Seeks opportunities to lead; in leading is attentive to each member of the team, articulates outcomes for each session and each project, keeps team on schedule, foregrounds collaboration and integration of individual efforts	Is willing to lead; in leading is attentive to each member of the team, articulates general direction for each session and each project, attempts to keep team on schedule	Will take lead if group insists; not good at being attentive to each member of the team, sometimes articulates direction for sessions, has some trouble keeping team on schedule	Resists taking on leadership role; in leading allows uneven contributions from team members, is unclear about outcomes or direction, does not make plans for sessions or projects	May volunteer to lead but does not follow through; misses team sessions, does not address outcomes or direction for sessions or projects, team members become anarchial	Never shows up and never contributes

1. Describe any communication problems within your group, or describe how well members of your group were able to communicate with each other.

2. Did you meet outside of class to establish goals and stay in tune with each other?

3. What worries you most when working in groups?

4. Did you think you did your fair share?

5. Did others do their fair share?

PEER AND SELF EVALUATION RUBRIC

Project/Activity: ____

Please rate your contribution to the group and evaluate each individual group member using a scale of 1–5, with 5 being the highest. Explain your reasons for any assigned ranking less than 4. Make sure you evaluate yourself on the first line.

Ranking   **Group Member**

______   ____________________________________

______   ____________________________________

______   ____________________________________

______   ____________________________________

______   ____________________________________

______   ____________________________________

Explanations, as needed:

Glossary

Alpha level The Type I error rate (false positive rate) used to test a hypothesis. It should be preset before data are collected.

Alternative hypothesis Formal statement that the treatment is effective.

Association Two variables measured on the same individual are associated if some values of one variable tend to occur more often with some values of the second variable than with other values of that variable.

Bayes' Rule A useful formula relating conditional and unconditional probabilities.

Confidence interval The statistic plus or minus the MOE; will depend on many things, including the level of confidence.

Confounding When the effects of two variables on the response can't be distinguished from one another (often caused by a lurking variable).

Correlation Measure of straight line association.

Correlation coefficient A common numerical measure of straight line association, always between -1 and 1.

Cross-sectional sample Non-random sample chosen to try and reflect a cross section of the population on features you deem important (e.g., race, income, education).

DISE Acronym used to help you apply the concepts of hypothesis testing.

Empirical rule A rule applying to bell-shaped distributions that contains useful probabilistic information about the variable being described.

Experiment Test under controlled conditions with the goal of making cause and effect statements.

Explanatory variable A variable we think might cause changes in the response variable. It's what you vary in your experiment.

False negative When a screening test says you don't have what the test is looking for, but you really do.

False positive When a screening test says you do have what the test is looking for, but you really don't.

H_A The alternative hypothesis (think "positive outcome").

H_0 The null hypothesis (think "negative outcome").

Human inference An informal way of describing the reflexive inferences we make when we consume simple statistical constructs, like charts, graphs and numerical summaries.

Hypothesis testing In statistical science, the paradigm for choosing between "Treatment Is Not Effective" and "Treatment Is Effective."

Inference The act of drawing conclusions about something we can't know for sure, based on information we have that we assume to be accurate.

Influence point An outlier such that, when removed, causes a distinct change in the correlation coefficient.

Lurking variable A variable not directly studied that can compromise your ability to attribute any changes in the response to a treatment. Often the cause of confounding.

Margin of error (MOE) A numerical way of acknowledging that you know your sample statistic is not going to give perfect knowledge about the population parameter; inextricably attached to some notion of confidence level.

Mean The arithmetic average of a set of numbers. Add them all up and divide by how many numbers you have.

Median The middle raw score. Arrange your data from smallest to largest. If you have an odd number of data points the median is the one in the middle. If you have an even number of data points then the median is the mean of the two in the middle.

Negative association When scatterplot is downwards to the right.

Negative predictive value The probability that you really don't have the condition, given the screening test says you don't.

Non-sampling error An error caused by something other than the fact that a sample was selected instead of the entire population.

Null hypothesis Formal statement that the treatment is not effective.

Outlier In a scatterplot, outliers are data pairs that are not spatially close to the bulk of the data.

***p*-value** The probability of seeing data that are as inconsistent with the null hypothesis, or more so, than the data you have just produced, assuming the null is true.

Parameter A number that describes the population characteristic you are most interested in.

Placebo effect The tendency for patients to show a real response to any treatment, even if it is inactive.

Population Larger collection of subjects or items that you are interested in understanding, which is way too large or too complex for you to examine each subject or item individually.

Positive association When scatterplot is upwards to the right.

Positive predictive value The probability that you really do have the condition, given the screening test says you do.

Practical significance Whether observed experimental differences are big enough to practically be worth caring about. Often called "clinical significance" in medical settings.

Prevalence The probability of the condition of interest appearing within the population.

Quasi experiments Studies that are unable to use randomization to evaluate effectiveness of interventions.

Randomization An action that produces groups of units that should be similar in all aspects (or eliminates potential biases due to the order in which treatments are administered).

Response substitution Process whereby some respondents frame their answers to questions in a survey in such a way as to express their views on issues outside the survey's scope—issues on which they have strong opinions.

Response variable A variable that measures the outcome of a study. It is the primary variable you are taking measurements on for your experiment.

Sample Collection of subjects or items that you select from the population to examine.

Sampling distribution The pattern of variability that a statistic exhibits when repeated sampling is done.

Sampling variability The variability seen in a statistic when repeated sampling is performed.

Scatterplot A simple *x-y* plot of two measurements (*x* and *y*) taken on each of several different subjects.

Sensitivity The ability of the test to correctly identify positive outcomes as positive outcomes. Numerically it is 1 − (false negative rate).

Simple random sample (SRS) A sample of n individuals chosen from the population in such a way that every set of n individuals had the same chance of being chosen.

Simpson's paradox When associations seen in a table change when that table is broken down into sub-tables.

Specificity The ability of the test to correctly identify negative outcomes as negative outcomes. Numerically it is 1 − (false positive rate).

Standard deviation The most common way of summarizing the spread or variability in a set of data.

Statistic A number of interest to the researcher that describes the sample.

Statistical significance When differences in treatments are sufficiently large that they are unlikely to be due only to chance. Formally, when a computed *p*-value is smaller than the preset alpha level.

Subjects Who or what the experiment is being done on.

Treatment What is being done to the subjects (for example, nothing or gastric freezing).

Type I error When the results from a hypothesis test suggest H_A is true when H_0 really is. A "wrong" positive.

Type II error When the results from a hypothesis test suggest H_A is false when it really isn't. A "wrong" negative.

Variance The square of the standard deviation. Common way to measure how spread out a data set is.